CONTENTS

CONSERVATION

1 THE ISSUES

At last humans are becoming aware that we are putting our Earth under increasing pressure. In the early 1950s a few far-sighted people began to worry that our activities were causing serious damage to the Earth. Now, after nearly 50 years, we are only just beginning to realise that the Earth is in trouble, and that if we do not take serious steps to halt the damage, our children will not have a viable future on the Earth.

Here is a more detailed description of the five main environmental issues that this book looks at:

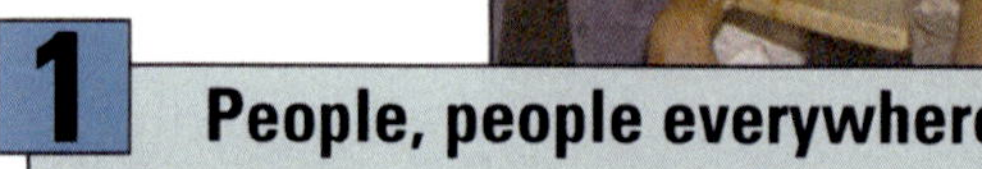

1 People, people everywhere

Humans depend on a healthy Earth's environment in order to survive. Until recently the Earth coped with our demands on it without being seriously affected. But the incredible population explosion of humans that has been taking place in the last 100 years has meant that the Earth cannot supply the needs of so many people in more and more places. Unless the human population growth levels off very soon, many parts of the Earth will be damaged beyond repair.

2 Pressure on living things

People use living things in order to survive – we need timber for shelter, plants and animals for food, and so on. The more humans there are, the more we need. This is being made worse by the fact that as humans become wealthier they want more from the environment. If too much pressure is put on the environment it cannot sustain itself – rich ecosystems are degraded into poorer ones, species die out and the Earth's life becomes poorer.

3 Pressure on resources

More people, and wealthier people, together mean greater demands for resources – more water is needed, more fuel for power, and so on. Many of the resources we want are already being used at a rate that cannot be sustained and many others are not renewable – once used they cannot be replaced. We have to learn to ration our resources. This is particularly a problem for wealthier countries. *The richest 20 per cent of the world's people use up 86 per cent of the Earth's resources.*

4 Waste

The more resources humans use the more waste we produce. Problems arise over what to do with this waste. It pollutes our air, our water and land. This is a special problem for the wealthier countries, because they also cause most of Earth's pollution and waste problems.

5 The future

Already there are signs that our demands on the Earth are causing changes to it. Global warming may be partly our fault. Humans must very quickly learn to look after our Earth better. We must learn to slow population growth, to protect and manage our environment, sustain and ration our resources and reduce and reuse our wastes. Or else!

2 THE TERMS

Look at these headlines about some of the important environmental issues of the new millennium. They all tend to use new and long terms like biodiversity, sustainability, ecotourism, and so on. These terms were not in use as recently as 10 years ago. What do they mean? We need to understand them before we look at the issues in detail.

The terms: what do they mean?

Biodegradable – able to decay into harmless chemicals by natural means. Some pesticides such as DDT, which used to be used widely, do not decay and they tend to build up in the bodies of animals in the food chain.

Biodiversity – the number of species of living things present in a community. Human activities are causing many species to become extinct, and the Earth's biodiversity is being reduced.

Biosecurity – the systems used by a country to prevent the entry of unwanted plants or animals. Often foreign species tend to take over local species.

Ecotourism – activities designed for tourists who are interested in seeing wildlife and experiencing it in its natural environment. Good ecotourism should not harm the environment.

Global warming – gradual increase in the temperature of the Earth due to the trapping of warm air by 'greenhouse' gases.

Population explosion – the increasingly fast growth of a population when controls that keep it in check are removed. Refers here to the human population after infectious diseases have been controlled.

Recycle – use again. Refers to waste that can be sorted, perhaps processed and used again in some way.

Sustainability – the use of natural resources at such a rate that no more are used up than are being replaced. For example, catching fish at the same rate that they produce young.

(There is also a glossary of other technical terms on page 48.)

PEOPLE, PEOPLE EVERYWHERE

3 THE SCARY GRAPH

The most important reason by far for the Earth being under pressure is people – billions of them! In fact we are living in the middle of an explosion of people! This is happening so fast that the Earth is having trouble adjusting. Its resources are being damaged or destroyed, and wastes from human activities are causing serious pollution.

Many people are saying that the Earth is in real trouble. But is this really the case, or are the problems being exaggerated by emotional environmentalists? To answer this we need to look at the evidence – the big graph on this page. Then we need to interpret it.

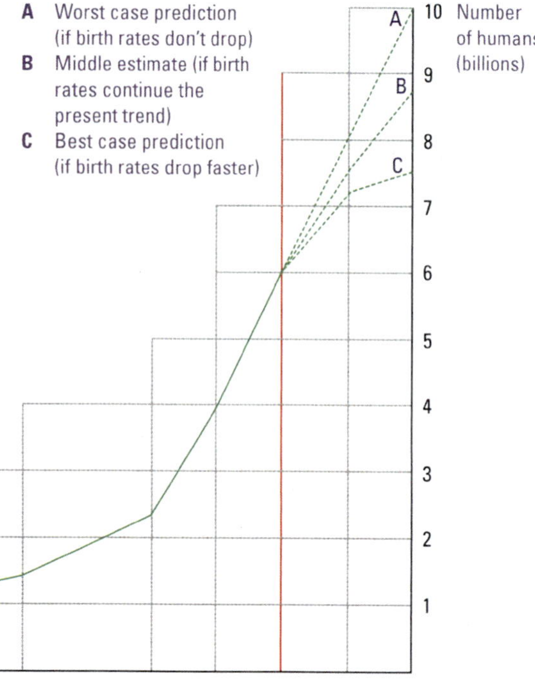

Graph of growth of world population of humans

Activities: Making sense of population statistics

1 Interpreting the population graph

Answer the following from the graph:

- a What is the total world population now; 100 years ago; 200 years ago?
- b What was happening to the world population numbers between 1600 to 1900?
- c What is happening to them during your lifetime?
- d In 0 AD the world population was 600,000.
  - How long did it take to reach 1 billion?
  - How long to rise from 1 to 2 billion;
  - from 2 to 3 billion;
  - from 3 to 4 billion;
  - from 4 to 5 billion;
  - from 5 to 6 billion?
- e Write a sentence summarising the recent trend in world population growth.

2 Predicting the future trends

The United Nations predicts future trends by using data from all countries.

- a One of the main aids in predicting is the average birth rate for a country. Why would this be important?
- b From the big graph, what is the predicted population for 2025 and for 2050?
- c What change can you see in the trend for each 25 year period?
- d In 1998 the UN predicted the world population in 2050 would be 9.4 billion. A year later they changed the prediction to 8.9 billion. In what way is this good news?

3 Drawing more detailed graphs

A closer look at population data shows that trends differ according to country. Generally there are two patterns – one in developed (wealthier) and the other in developing (poorer) countries.

- a Draw a graph to show the patterns in these countries as examples. Set your graph out as for the world graph. On the vertical axis mark out to 12 million people.

	Developing country	Developed country
Year	Zimbabwe	New Zealand
1950	2.850 million	1.910 million
1970	5.515	2.830
1990	10.100	3.360
2000	11.340	3.820
2020	10.000	4.550
2050	9.280	4.840

- b For each country describe the shape of the graph line up to the year 2000.
- c Explain how the graphs differ.
- d Which country has the higher birth rate?
- e What is predicted to happen to the population of each country in the next 50 years?

Why has the population of the world exploded?

You will have noticed that the world's population began increasing rapidly over a relatively short time – the first half of last century. What happened during this time to cause this? It is all to do with the scientific discoveries of the later part of the nineteenth century, particularly to do with health. You will also see that the effects differed between wealthy, developed and poor, developing countries. Try the activity below and you will see how these life-saving discoveries are to blame.

Activity: Explaining why the population of the world has exploded

The flow diagram summarises some of the reasons why the population of the world has been growing so fast. It also shows why it is growing fast in developing countries and not in developed ones. The reasons are jumbled up in the boxes.

1 Draw the flow diagram. Then use the colour coding to choose the reason that fits each of the circles. Write the number of each reason in the correct circle. The reasons must be in the correct order.

2 When you have finished, compare your answers with others in your group. In some cases you may be able to find good reasons for having a different order.

World Population Growth Flow Diagram

THE POPULATION GREW VERY SLOWLY.

POPULATION GROWTH RATE IS FASTER THAN THE DEATH RATE.

PEOPLE ARE WEALTHY AND POPULATION GROWTH IS SLOW.

OVERCROWDING AND STARVATION ARE COMMON AND POPULATION CONTINUES TO GROW.

1 More food and fuel are needed to feed the extra people.

2 People are well-educated and have money to buy many possessions.

3 Both death and birth rates are low.

4 Death from disease or hunger almost equalled births.

5 The link with bacteria and disease was discovered.

6 Fewer children tend to mean a more comfortable lifestyle.

7 Fewer trees mean less rain so droughts occur.

8 People are poor and have little education.

9 People had to work hard to survive.

10 Birth control methods widely used.

11 Country people move to the cities.

12 Hygienic health measures were discovered.

13 Fewer children die as improved health measures come from the wealthy developed countries.

14 Educated people began to understand science.

15 For poor families children are seen as security for the future.

16 Fewer children died of disease and people lived longer.

17 Land is over-cropped and over-grazed and forests are cut down for fuel.

4 The have-nots

The population explosion has affected the poor countries most severely. More people means more pressure on the land for more food. In drier countries crops fail more often because of more droughts. More droughts are partly because more people mean more demand for firewood and so more trees are cut down. Fewer trees mean less moisture in the air and less rain. This pattern of repeated famine is happening most often in parts of north-east Africa – especially Ethiopia.

Famine in Ethiopia

In the year 2000 the rains had failed to come to southern Ethiopia yet again, and over 12 million people were facing severe food shortages. The news item tells how desperate the nomadic herders in the worst hit areas had become. These people could only be helped by massive aid from wealthier countries overseas. But:

1 The Ethiopian government was spending most of its money fighting a war with a neighbouring country (over land needed for the increasing population).
2 Some countries were wary of sending food to a country where the government was at war.
3 It was very difficult to get food aid to the starving people. Roads were terrible, and most trucks had been sent to the war.

So overpopulation, drought and lack of food put these people in a hopeless situation.

A barren land that grows only sorrow

DANAN – As the desert sun began to dip towards the scorching horizon, Bashir Ahmed Abdi died.

The 3-year-old had spent his last day unconscious in a tiny hut built from sticks and sacking, in the middle of a barren, red-earthed plain. His mother fed him a little water around noon. It was all there was.

When she realised her youngest son was dead she slowly pulled a faded blanket over his body. He was the third of their five children to die …

The family's hut is at the edge of a settlement of nine thousand starving people that did not exist a month ago. Now it swells every day as more of southern Ethiopia's desert farmers stagger in from the bush.

Danan, a scrappy village in normal times, is on a road, and a road means the chance of food. So the ragged walkers keep arriving.

Some, like Bashir's family, have trekked 240 km, mainly at night to avoid the heat. Five months ago Bashir's father had 40 cattle and 50 sheep and goats. They had provided milk and meat and money for other things. Now the livestock have all died, so the people had to abandon their villages.

In Danan the little food that aid workers have been able to provide has nearly run out. Eight children are dying every day. There are no toddlers or old people to be seen.

Even if aid does arrive soon, 80 per cent of the region's sheep and cattle are already dead. The next rains are not due for six weeks – if they come at all. Many farmers are talking of going to the cities – if they survive.

Modified from a report in *NZ Herald* of 12 April 2000

Activity: Looking at the difficulties of Ethiopia

1 Analysing the situation in Ethiopia

a From your atlas draw a map of Africa and mark in Ethiopia. From a world rainfall map note the approximate amount of rain southern Ethiopia gets each year. How much does your district get?

b Draw a population graph for Ethiopia using this data. (Set it up as in page 4, question 3.)

1950	20.1 million
1970	29.7
1990	48.3
2000	64.1
2050	185 (estimate)

c Summarise what the graph tells you about the growth rate.

d From the estimate do you think that Ethiopia is likely to have more famine problems in the future? Why?

2 Interpreting the newspaper report

Read the report and answer the following:

a How do the people normally live and get their food?

b Why did they leave their land and go to Danan?

c How many children did the family have?

d Why were there few toddlers and no old people in Danan?

3 Sorting out the problem

Discuss in your groups:

a How the repeated droughts and food shortages are related to the population problem.

b Why there has not been much aid from overseas countries so far.

c Why the people cut down trees even though they attract rain.

d What responsibility (if any) the following have for the Ethiopian famine problem:

The Ethiopian government, the local people, the United Nations, the wealthy countries, New Zealand.

5 Moving to the cities

All round the world there is a vast movement of people taking place. They are migrating from the countryside into the cities. It is happening in the rich countries, but much more in the poor ones. In China this move of people from rural areas to cities is the greatest mass migration of people in history. (See page 10 for more detail on the growth of cities.)

Why move to the cities?

Why does a poor Ethiopian herder, or a boy from an overcrowded village in Bangladesh move to the city?

1. *Pressure on the land* – more people means that the worn-out land has to produce more food. In many places it cannot.
2. *Lack of work and money* – young people from the countryside come to the cities in search of jobs, money and a better life.

Urban slums

Unfortunately the new cities often do not provide enough work, and the new arrivals finish up living in vast city slums. They are places where people struggle to survive with little work and dreadful crowded living conditions, often much worse than in Langa. There may be no toilets and any water supplies may need to be shared by many families, and are very likely to be polluted. As a result health problems caused by overcrowding and food shortages are common. It sounds terrible. *Yet three of every ten of the world's children are being born into extreme poverty like this.*

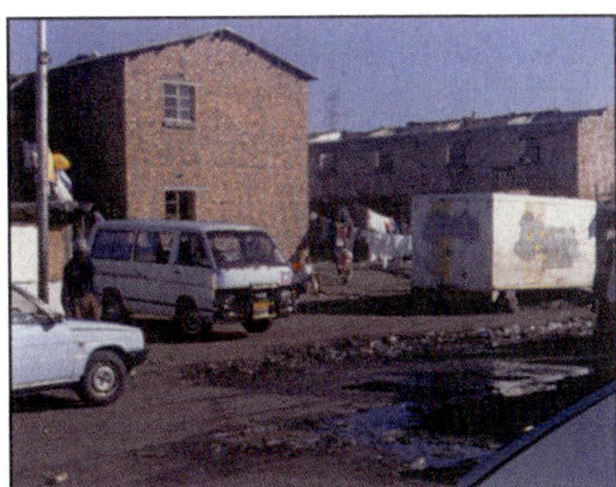

The people of Langa live in old brick hostels – and overflow into the packing cases and old containers. Most life is on the street.

These children are looking forward to dinner – a sheep's head being cooked in a drum.

Living in Langa, a 'Township' on the edge of Capetown

On the fringe of Capetown, in South Africa, there are several 'townships' which are home to over a million people who have moved from the countryside to the city looking for work. The oldest of these is Langa, where most people live in old brick buildings that were built more than 30 years ago as hostels for men. A typical hostel has a central shared space for cooking and eating. Leading off this are three or four small rooms. Each room is about 15 square metres, has space for three beds and is the home for three families! Imagine three families of perhaps four or more people sharing this space. There is one bed per family. Their clothes and any other belongings have to be hung on the walls.

The families spill outside into lean-tos and old containers. Most cooking is done outside on open fires in old drums. Water supplies and toilets are shared by many families. There is mud and litter everywhere. Few people have regular work, and there is not enough money in the township to collect the litter or improve the muddy roads – yet Langa is one of the better townships.

The newest arrivals settle in the new township of Khayelitsha. Here conditions are much worse. Most homes are made of cardboard, packing cases, sheets of tin, anything that can be found. Most have no water, electricity or toilets.

Activities: Finding out about life in a slum

1 Life in Langa

In your groups discuss the living problems of the people of Langa, applying them to your house:

a. If 10 people slept in each bedroom in your house, how many people would live in it? What would it be like?
b. Organising the cooking arrangements for these people.
c. Sharing one portaloo and one cold water tap.

2 What do the photos tell us?

Look at the photos. What give you clues to the following:

a. The township is overcrowded.
b. There is little work for people to do.
c. The Langa council has little money to improve conditions (two reasons).
d. The people are very poor.
e. There is a lack of cooking facilities.

3 Moving to the city

a. List three reasons why a young person might want to move to the city.
b. List three problems he or she is likely to have to deal with to survive in the city.
c. List three reasons why many of the ghetto children die of infectious diseases or malnutrition.

Maggie is one of the wealthier people in Langa. She owns the house behind her, and has her own portaloo.

6 Defusing the population bomb

Can the population explosion be stopped?

The graph on page 4 shows population growth out of control. There are 247 babies being born every minute. This is equivalent to the total population of a city like Christchurch or Wellington being born every day (over 350,000). This rate of growth is faster than our ability to feed all the world's people. Nearly all of this growth is in the poorer developing countries while it has slowed in the wealthier countries. Can this also happen in the poorer countries before we have widespread famine and disease?

The problem of the next 50 years

The latest United Nations predictions suggest that the rate of growth of the world population is slowing very slightly. (Check back to the graph on page 4.) Even so, it predicts that in 2050 the world population will be 8.9 billion, and will probably peak at nearly 10 billion. Some of the developed countries may even show a population decline.

This may seem good news, but can the world cope with a further increase of 3 billion in the next 50 years without widespread famine and disease? Already some places, such as Ethiopia, have more people than the country can support. Can we find better ways of getting food from the rich areas to those that need it? There are still huge problems to be solved in the next 50 years.

Data on birth rates

The birth rate or fertility rate is the average number of babies born per woman in her lifetime. It is a very important indicator about the population growth of a country. Here are some examples:

	Birth rate in		
	1950	2000	2025
Developed area			
A Europe	2.6	1.4	
Developing areas			
B Asia	5.9	2.6	
C Africa round the Sahara	6.5	5.5	
New Zealand	3.6	1.8	1.8
Ethiopia		7.1	4.9

Controlling growth

In the wealthier countries population growth has slowed because people have enough wealth and education to feel that having fewer children means a more comfortable lifestyle, and family planning has become important.
It is more difficult in poor countries. Two things can be done:

1 **Use birth control** (see page 9).
2 **Raise living standards** by improving the wealth of the countries.

Here are some of the ways that 'have' countries have tried to help. They have not always worked:

A Ways of helping poor countries

1 **Provide food aid** in an emergency such as a drought.
2 **Send experts** who can provide advice and help, for example by setting up local industries or building water wells.
3 **Help improve farming** by developing improved crops and farming methods.
4 **Provide loans** to the government for developing new industries.

B Reasons why help may not work

1 Improving farming (new seeds, fertilisers, etc.), is often so expensive that cash crops have to be grown to sell overseas rather than feed the people.
2 Giving extra food is only a short-term solution. It makes people dependent on other countries.
3 Money from loans sometimes finishes up in the hands of corrupt politicians and business people.
4 Setting up new industries, etc., often fails when the experts leave because the people do not have the skills to keep them working.
5 Money loans often leave huge debts to be paid off.

Activities: Dealing with the population problem

1 **Analysing the birth rate data**
a For each of the three areas A, B and C, how much has the birth rate dropped in 50 years?
b In which of the developing areas B and C has most been done to slow birth rates?
c What will happen if the average number of children is less than two per family?
d What do you predict for the population numbers in New Zealand and Ethiopia by 2025?

2 **What about death rates?**
The population of the world is also affected by how long we live. In most countries people are living longer.
a Does living longer slow down or speed up population growth? Explain.
b What must be happening in most countries to cause people to live longer?
c The average age of death in India is 62 years. In New Zealand it is 78. Explain the difference.

3 **Problems with helping poor countries (from the two boxes above)**
The four examples of ways of helping (Box A) all seem sensible. But they all have problems.
In your groups work out which of the five reasons why help may not work (Box B), matches the ways of helping in Box A. Discuss what can be done in each case.

Population control – getting the right answers

Many children are needed because most will die.

Male children are needed to help work the land and look after us when we are old.

The old ways are best. Birth control advice is a government attempt to change the old ways of life.

We don't like outsiders telling us what to do.

Contraceptives cost too much

The birth control message

Governments in developing countries have been trying to get the message across that fewer children are needed, since the United Nations created a population fund in 1969. At first when governments tried to do this they were met with suspicion and objections like those above. Governments have tried to do this in several ways. Some of these are:

A **Abortion** – some countries, such as Japan, made it easy for women to have abortions.

B **Target-based birth control** – India set up government family planning clinics that offered money or transistor radios to men or women who had a small operation to stop them having children.

C **The one-child policy** – China passed a law that only one child was allowed per family. Punishments include losing family benefits, homes or jobs.

D **Free contraceptives** – government family planning clinics put money into providing free contraceptives.

E **Education** – government family planning clinics give advice to people about the value of family planning, especially to women.

F **Public advertising campaigns** – including radio, television and films, which explained the benefits of fewer children. (Some, aimed at men, explained that it is more macho to clothe and feed your children than to just produce them!)

Activities: Answers to the birth control problems

Birth control posters in China

Discuss in your class or group:

1 Understanding different viewpoints

For each of the birth control programmes above:

a Work out reasons why poor, uneducated village people might have objected.

b Discuss the pluses and minuses of each programme.

2 China's solution

Method C has been very successful. It has reduced the growth rate of China's huge population of 1.2 billion people from over 2 per cent to less than 1 per cent in the last 30 years. But there are problems. Discuss these things:

a Male children are treasured and spoilt. They are seen as more valuable because they can support their parents when they get old.

b There are many more male children because female unborn babies are often aborted or left to die.

c How successful this law would be in New Zealand.

3 India's solutions

a India first tried method B. This was very unpopular and some officials put in false records. It failed – why?

b India is now having success using E and F. Why might this be?

c Indian health workers now believe the best method of birth control is to educate the women. Explain.

4 Family planning in New Zealand

Discuss in your class or group:

a If family planning programmes are needed in developed countries like New Zealand.

b For each of the programmes listed, if it would be acceptable in New Zealand.

c The rights of individuals in family planning.

7 The growth of cities

All around the world there is a huge migration of people into cities. In 1900 10 per cent of the people lived in cities. In 2000 nearly half of the world lives in cities. The biggest cities, 25 years ago, were places like **Tokyo**, **New York**, **Paris** and **London** – mostly old cities of the developed world.

Most growth now is in new cities in less developed countries. The UN estimates that by 2015 the five largest cities in the world will be **Tokyo**, **Mumbai** (India), **Lagos** (Nigeria), **Sao Paulo** (Brazil) and **Dhaka** (Bangladesh). These mega cities will each have over 20 million people yet, apart from Tokyo, they are all in developing countries.

A These huge tower blocks of small apartments in Poprad (Slovakia) are typical of how the communist countries solved the housing problem caused by the move to the cities in the 1950s and 60s. Often these places become slums that are unpleasant to live in. Why?

The problems of uncontrolled city growth

You have seen on page 7 why people are moving to the cities, and how in the poorer countries this results in huge urban slums with environmental problems. Unfortunately, city growth in wealthier countries is also causing environmental problems that are just as serious. Some facts about the world's cities:

1 600 million of the world's city dwellers have inadequate housing.
2 In China alone 1 million die from air pollution every year.
3 Cities use 60 per cent of world water resources (and cover only 2 per cent of the land).
4 Traffic delays cost business about $1 billion per year in Auckland alone.

Phew! Why would anyone live in a city?

B Botany Downs, Howick. Modern cities in wealthy countries tend to solve their housing problems by urban sprawl – letting new suburbs spread out into the surrounding farmland. Why might this be a problem?

Activities: Cities and their problems

1 Locating the cities

a On a world map shade in green the old developed part of the world. Find and mark in the four biggest cities of 25 years ago (mentioned in the text above).

b Find and mark in the five cities that are becoming the largest.

c Find and mark in Capetown (Langa, page 7, is a suburb).

2 Comment on the quote

Cities are... 'devouring water, food, energy and processed goods, and then belching out the remains as pollutants...'
Do you think that in general this comment is true? Why?

3 Why people move to the city

Draw a diagram to show the reasons: either why people move to cities to live or if your family has moved to live in a city (either from the country or perhaps a Pacific Island), why they did this.

4 Analysing some solutions

Look at the two photos of city housing. For each:

a List ways they help solve the housing problem.

b Describe the pluses and minuses of growing up in such places.

5 Who would live in cities?

From the above list of four facts about growing cities, put down the numbers of the problems that are evidence that:

a Urban sprawl is using up good land.

b Cities use up resources of water, food.

c Cities are major polluters.

d Cities have problems providing shelter.

e Cities have major traffic problems.

6 Publicity about migrating to the city

Design a poster showing the things one needs to know about and be prepared for, before moving to live in a big city from the country.

Making cities better places

Although people everywhere are moving to cities, even in wealthy countries cities are not always good places to live in. Often the organisations who run the cities struggle to provide a good **infrastructure** (see box). If there is not enough money, or planning is bad, then the lifestyle of the people will be poor, as in Langa.

Modern cities try to plan ahead to solve problems. These photos show some of the problems (and solutions) of cities around the world.

The problems of cities

These are the things (infrastructure) that cities need to provide for to make them good places to live in.

Providing the infrastructure:

- Housing and sprawl
- Good transport
- Water supplies
- Power supplies
- Waste removal (landfills)
- Storm water disposal
- Waste water/sewage disposal
- Air pollution control
- School/open space provision
- Disagreements amongst local bodies

Lifestyle problems:

- Crime in crowded places
- Isolation (living in suburbs)

C Commuters in Beijing (China). As their lifestyles improve many of these people want cars. How will Beijing cope?

D Sometimes the air in Delhi (India) is so polluted that traffic police wear gas masks. What would need to be done to reduce air pollution?

E This Auckland reservoir is full, but in 1998 the reservoirs were running very low and water had to be rationed. What must be done to prevent this happening again?

F This housing is in a poor city in India. The people living here would have several problems. What are they? Which problem would you find the most difficult to live with?

G These power pylons were put up in two weeks to solve a power blackout from broken cables in Auckland. What problem does this highlight?

H Look at the photo of the city rubbish tip in Cambodia on page 38. List the problems of having this sort of rubbish dump.

Activity

For each photo and also the two photos on page 10 and one on page 38:

- a Identify the problem (from the box above) that each illustrates.
- b State if it shows a problem or a way of solving the problem.
- c Answer the question that goes with that photo.

8 AUCKLAND'S TRANSPORT

An example of the problems of a growing city

Every day in Auckland many of the nearly 700,000 car and truck drivers get angry. They get angry because they get stuck in traffic. Many are spending over 40 minutes getting to and from work each day. And it is rapidly getting worse. Each year the traffic increases by 5 per cent. Average speed has dropped on one motorway from 68 to 34 km/h in six years.

Auckland, with its population of 1.2 million, is by far the biggest city in New Zealand, and its traffic problems are a good example of the sort that develop in cities that are growing fast. The traffic problems have developed for many reasons, but mostly because in the past the city grew without clear planning.

- Urban sprawl was unchecked, so the city is very spread out.
- Public transport plans of the 1970s were thought to be too expensive. As a result bus and rail services are poor.
- Aucklanders became used to using cars and now 87 per cent of trips are by car. In the 1960s motorways built to solve the problem just encouraged more car use.

Driving us crazy

Four in the afternoon on the southern motorway heading north. Motorists cruising at 100 km/h around Tip Top corner suddenly hit the brakes.

In front of them is a 10 km tailback. The rest of their journey will be a bumper-to-bumper, stop-start affair, until they have negotiated the Harbour Bridge, or got off along the way.

The full crawl over the bridge might take half an hour... if all goes well. These conditions will continue for three hours.

All over Auckland, drivers are experiencing varying, and worsening, degrees of frustration in peak-time delays...

From *NZ Herald* 25 March 2000

Conflict about solving the problem

Now Aucklanders are desperate, and something has to be done. But what? The city planners are faced with conflict about how to improve traffic flow. Any solution will be expensive and there are different views.

A More motorways first priority

This view is being pushed by the Business Associations, who want up to $1.6 billion spent on new link motorways. Their reasons include:

1 The traffic problems cost millions of dollars in lost business.
2 Most people already rely on cars and do not want to leave them.
3 Public transport would not clear the roads enough.
4 The city is too spread out and trips are too varied to be solved by public transport.
5 Using rail for freight (e.g. containers) is too expensive because it is inflexible and involves double handling.

B Better public transport first priority

This view is being pushed by people who want to see a variety of public transport alternatives such as: More buses and busways, better rail transport, a rapid rail system, cycleways, ferries.

Their reasons include:

1 More motorways are not long-term solutions. They encourage more car use, and become full in a few years.
2 Good public transport shifts more people faster, is much less polluting and uses less energy.
3 Better public transport gives people alternatives to cars.
4 Local communities do not want motorways through their parks and reserves.

Traffic on Auckland Harbour Bridge.

The city planners' solutions

The Auckland Regional Authority and City Councils have problems in agreeing on the best solutions. However the general view is:

1 **Public transport:** Several solutions are being considered. (Together they would cost over $1 billion!)

- Busways – special bus lanes from North Shore over the bridge, costing $ 130 million, are already being worked on. Some other bus lanes are already in use.
- Improving passenger rail – better stations and service, with a rail connection to the heart of the city.
- Rapid rail – there are plans for a loop rapid rail round the central part of the city.
- Cycleways and more ferry services on the harbour.

2 **Motorways:** Recent councils are spending most of the transport money on a mixture of public transport and the completion of some motorways – especially on major motorway links.

Activities: Solving Auckland's traffic problems

The city planners have difficult decisions to make over transport. In your group imagine you are the Council members. You must deal with the following:

1 Making decisions
For each reason given by both the motorway and public transport supporters, decide:
a Is the reason strong or not.
b Which solution you as councillors would make your first spending priority.

2 Deciding who should pay
How should the money for paying for improved transport be obtained? For each of these state if they should pay and why:
a City ratepayers
b Government (from general taxes)
c Heavy truck users (they already pay a road-user tax)
d Car users (petrol tax/tolls on roads).

3 Public transport options
For each of the public transport options mentioned, list its pluses and minuses.

4 Different viewpoints
In your groups consider each of the following:
a Explain their viewpoint. Is it reasonable?
b Explain how you think the Council should respond.

A Commuter: 'I have always used my car to go to work. It is more convenient than a bus.'

B Commuter: 'I like to go to work in my car and I don't want to be told to give it up. More and cheaper city car parks are the answer.'

C Railways manager: 'We don't make a profit from passengers because not enough people use rail, so it is not worth improving the service.'

D Businessman: 'Clogged roads cost me money because they slow the trucks which bring containers from the wharves to my workplace.'

E Mother: 'I have a family and I am concerned that pollution from city traffic will harm my children.'

F Home owner: 'I live near a nature reserve and I don't want a proposed motorway to go through the reserve.'

G Bus company manager: 'I cannot afford to put buses on new routes without subsidies.'

H Home owner: 'If a motorway is put near my home its value will be reduced.'

Activities: Local issues

1 New Zealand's population growth

New Zealand population data		
Year	Total	Birth rate
1950	1.910 million	3.6
1970	2.830	3.2
1990	3.360	2.2
2000	3.820	1.8
2020	4.550	1.8 (est.)
2050	4.840	?

a Plot total population data on a graph (Set it out as for page 4, question 3.)
b Describe the growth pattern. Is New Zealand's a developed or developing pattern?
c On a second graph plot the birth rate. Explain how this has changed.
d The age of death is rising slowly. What effect will this have on population growth in the future? What problems will this cause?

2 Population patterns in my class
Each class member find out (if you can) from your parents/grandparents:
- how many children were in the families of your parents/ grandparents/earlier ancestors.
- where you were born. Also where your parents and grandparents were born.

Discuss this information in class.
A *Family sizes.*
From the class information:
a How have family sizes changed over generations?
b Is this the same as the global trend?
B *Migration.*
From the class information:
a What proportion of the class has migrated to live in another place in your lifetime?
b Have the parents/grandparents tended to move to other places to live in their lifetimes?
c If you are in a city school what proportion of your families have moved from the country to live in the city?
d What proportion have moved from the city to the country?
e If you are an Auckland school, what proportion of families have moved to Auckland to live?

3 The problems of my town
The 70 per cent of you that live in a city choose that city. Country students choose your nearest large town. Do some research about your town. Find out:
a About its population:
- How big it is
- How fast it is growing (or declining)
- The ethnic mix.

(See the references on NZ population statistics on the inside back cover.)
b What growth problems it has. (Check the box at the top of page 11 for a list of possible problems. Each city may have had to face special problems – for example, winter smog in Christchurch, sewage disposal in Wellington. Auckland, because it is larger and faster growth has many problems apart from transport, for example, power, water supply, sprawl.)

Choose one problem and find out more about it:
- What the problem is.
- Why it has developed.
- What is being done about it.

Present your findings as a report or poster.

Living Things Under Pressure

9 Shrinking Rainforests

Although tropical rainforests cover less than 6 per cent of Earth's land surface, they are the richest and most productive habitats on Earth, and they are disappearing fast. Why is this happening and why are they so important?

What is happening?

The tropical rainforests are rapidly disappearing. In the last 80 years over half of the world's rainforests have been destroyed. There are less than 9 million square km left.

- The rate of destruction is so fast that every year an area larger than the North Island is destroyed (150,000 square km). This is equivalent to two football fields every second!
- It has been calculated that as many as 135 species of forest plants and animals are becoming extinct every day – a huge loss of biodiversity!
- The rate of destruction is not slowing down.
- In dry seasons smoke from fires set to clear the forest may be so thick that planes cannot fly and people become ill.
- When the forest is removed droughts become more common and the soil quickly loses its fertility.

Why are the rainforests important?

- **Variety of living things:**

Rainforests are made up of a huge variety of trees, shrubs and other plants and animals. In fact they make up half of the species of living things on the planet. This is a vital source of species that may be useful in the future.

Rainforest in Ecuador, South America. Note the wide variety of plants. It is very humid and wet, with many birds living in the canopy, and insects and small mammals lower down. Up to 5000 different kinds of living things may live in one tree!

- **World climate effects:**

Because of the huge amount of water that evaporates from the rainforests every day, they are the source of much of the rain that falls in the tropics.

- **Oxygen release:**

The enormous amounts of oxygen released by plants in photosynthesis are vital for animals to breathe.

- **Carbon dioxide absorption:**

The rainforests absorb huge amounts of carbon dioxide from the air every day, preventing its build up.

- **Home for native people:**

The Amazon forests alone are still the home for over 200,000 Indians. Tropical rainforests were once the home for many millions of native people.

Why are the rainforests being cleared?

There are three main reasons:

1. **Population growth** – This is so fast in most countries with rainforests that people need more land for farming.
2. **Timber milling** – Large timber companies lease the forests to mill the valuable hardwood timber. Usually they have no interest in replanting.
3. **Commercial farming** – Large-scale cattle ranches or plantations are set up to make money for overseas markets.

Activity: Getting to know the rainforest

1 Locating the rainforests

a From your atlas find and name the three main rainforest areas (A, B, C on map).

b Where is the nearest rainforest to New Zealand?

c From your atlas look at a world rainfall map. What is the rainfall pattern in the rainforest areas?

2 How long will the forest last?

Check above for the amount of forest left now and its present rate of removal. From this work out:

a How much will be left in 20 years?

b How long before it is all gone?

c How much will be left when you are 80 years old?

3 Why are the rainforests important?

a List three ways that the tropical rainforests affect the world's atmosphere and climate.

b List reasons why the biodiversity of the rainforest is important to scientists.

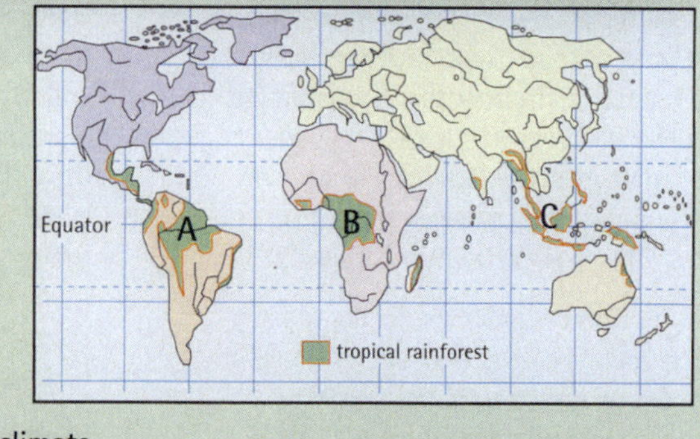

Activity: Conflict over rainforest resources

The removal of rainforests is a complicated problem and different people see it in different ways. This example shows some of the pressures on forest removal in the forests of Kalamantan in Indonesia.

1 These are some of the groups affected: Match each up with the number of the box that best describes how they are affected.

A	politicians	B	native people
C	scientists	D	commercial farmers
E	tourists	F	timber companies
G	small farmers	H	conservationists
I	overseas banks	J	businesspeople
K	airlines and people with breathing problems		

2 Make two lists:

A Four or more reasons why the forest removal should be stopped.

B Four or more pressures on government to let forest removal continue.

1 Study the forest life to learn how to manage it and to find plants and animals that may be useful in the future.

2 Clear forest to set up small farms to grow food. The soil quickly becomes poor and they often have to leave their land.

3 Lend money to Indonesian government and businesses for profit.

4 Mill the timber from forests leased from the government and sell it overseas.

5 Lease forest to timber companies and develop industries to sell goods overseas and repay overseas debt.

6 Live in the forests, hunting, gathering and selling a little wild rubber, etc.

7 Visit the forests to enjoy them and to see the culture of the local people.

8 Set up businesses to mine the resources and mill the timber to make money.

9 Set up large farms in the cleared forests to grow crops such as coffee and cocoa to sell overseas.

10 Protest at the destruction of the forest.

11 Are affected by smoke from fires from clearing fires that rage through the forests in times of drought.

Activity: The plight of the Secoya people

The Secoya Indians live in the Ecuadorian rainforests. Yanqui's tribe once lived a nomadic life as hunters and gatherers in the forests until they were discovered by missionaries about 60 years ago when he was a boy. His people now live in a small village several hours journey by boat and road from the small oil town of Lago Agrio. Much of the surrounding forest has been destroyed and is being farmed by people who have come into the area looking for land. Many of the roads that are opening up the forest for farming and timber milling have been put in by the government-owned oil company.

There is little work to support the young people in the village. Most leave the village to get work with the oil company or in the cities.

In small groups discuss if the following are good or bad for the Secoya people:

1 When Yanqui dies his culture will be gone because the young people are not interested.
2 A tourist company has a camp in the nearby jungle. It pays a little money to the Indians in return for allowing the tourists to see their way of life.
3 The missionaries made the people wear clothes and tried to make them Christians (many still believe in the shaman or 'witch doctor').
4 The oil company provides a little work for the local people and has put in roads, although its profits mainly go out of the region.
5 The Indians grow some food crops, but money is needed to buy clothes, some food and petrol for the motors on their dugout canoes.

This man is Yanqui, an elder of a small tribe of Secoya Indians living in the Ecuadorian rainforests. He tries to keep his culture alive by demonstrating to tourists how they used to use hunting blowpipes and other tools.

10 Shrinking New Zealand habitats

It is not just overseas areas that are suffering from the effects of humans on their natural ecosystems – communities of living things. New Zealand's natural communities have also been greatly changed since humans arrived.

Why are New Zealand's natural communities special?

For over 80 million years New Zealand has been an island, isolated from the rest of the world. Its communities of living things developed in special ways:

- All of the country was covered with native vegetation – mostly forest.
- There were no land mammals.
- The plants did not need to develop means of protection from browsing animals.
- Many birds lost the power of flight because they did not need to fly.

Because of this our native animals and plants have evolved some very special and unusual features. In fact, most species are endemic (different from those anywhere else in the world). The first human arrivals found a country that was mostly covered with forest (over 80 per cent). The rest was either alpine (mountains), scrub, tussock grassland or wetland.

The first human invaders – the Moa Hunters

It is thought that the first Maori arrived around 1000 years ago. They came from islands of the Pacific. They brought with them:

- A culture that relied on living off the local environment. They grew some food, but mostly were hunter-gatherers, relying on fish, birds, shellfish and seals.
- They also brought a dog, the kiore or native rat, some food plants such as kumara and, very important, they brought fire.

We are now beginning to understand that the early Maori had two very important effects on the natural environment:

- **Moa hunting:** The excavation of moa-hunter sites shows that they were hunted on a large scale (bones of up to 500,000 moa have been found). The result was that by 500 years ago the once common moa were extinct.
- **Burning of the forests:** Between 1500 and 1600 AD, huge areas of the forest were burned. In the eastern South Island, in particular, almost all the forest was burnt and replaced by tussock grassland. It is not clear why this happened – possibly fire was used to help hunt the moa and may have got out of control.

After about 800 years as much as one third of all the forest had been destroyed. About one quarter of all the land birds (34 species) had become extinct, and the populations of other larger birds and seals were much reduced.

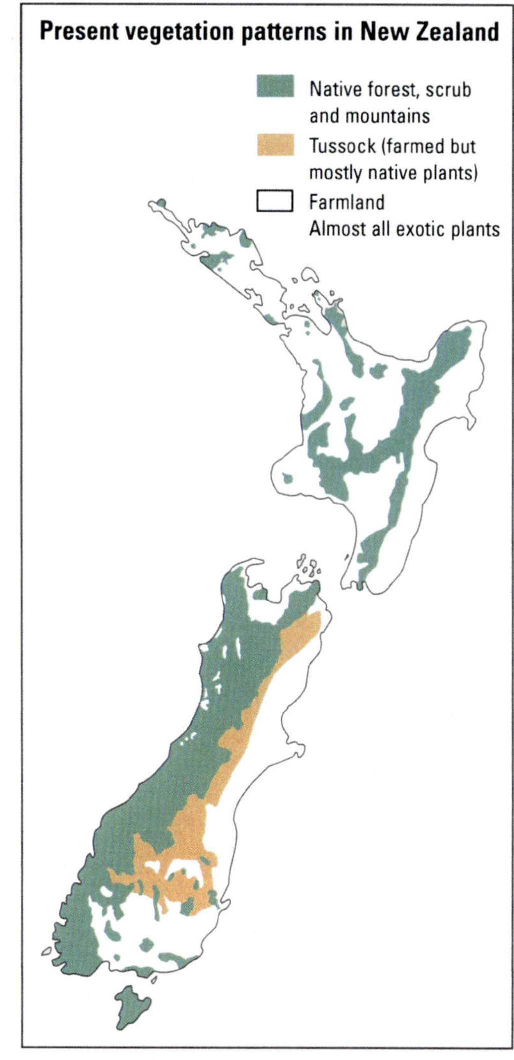

New Zealand vegetation types – percentages

	Pre-humans	Present day
Indigenous forest	81	21
Tussock	5	14
Alpine/scrub/etc.	14	19
Exotic forest	–	2
Farmland	–	44

Activities: Analysing the effect of humans on New Zealand environments

1 Terms
Write down definitions of these terms (See pages 14, 16 and 18): biodiversity, browsing, ecosystem, endemic, exotic, habitat, indigenous.

2 Vegetation changes
a From the table draw two pie diagrams of the pre-human and present day vegetation.
b Write sentences explaining the changes between the two.

3 The indigenous vegetation:
Explain why:
a Many of the birds were very easy to hunt.
b Many of the plants were more damaged by browsing than plants from other countries.

4 The effect of the moa hunters
Draw diagrams to show:
a The effects of the fires on the forests
b And on the flightless birds.

5 The effects of the early Europeans
For each of the three effects on the next page draw a diagram to summarise:
a Why the settlers did this.
b How it affected the living things and the landscape.

The second human invaders – the Europeans

In the last 200 years the impact of the second wave of human invaders on our environment has been even greater. It started in 1769 when Captain Cook released pigs and goats – and rats escaped. Then whalers and sealers arrived, and after 1840 immigrants came in large numbers. They were mostly from Europe – especially Britain – and they came looking for gold and for land to settle on and farm. In 1800 the population of New Zealand was about 100,000 Maori. By 2000 it was 3.6 million people. These arrivals have had huge impacts on the environment – affecting it in three main ways:

This forest is made up almost entirely of native plants.

1 Hunting: The early arrivals hunted and collected specimens of many native living things. Sealers and whalers almost wiped out the seals and whales.

2 Clearing land for farming: Timber milling destroyed large areas of forest. The settlers drained wetlands and burned off vast areas of forest and scrub so that by 1920 half of the remaining forests had been converted to farms and towns.

3 Introduction of plants and animals from overseas: When the settlers began to farm they brought out, mainly from Britain, all kinds of plants and animals that they were used to and needed.

New Zealand's landscapes now

If we look around New Zealand now most of the landscapes where people live are largely covered with plants from overseas – pasture of grass and clover, crops, forests of pine. Even most gardens are largely filled with plants from overseas. All the mammals we see are from overseas, as well as most of the birds and many of the insects. The remaining native communities are mostly in National Parks and remote places.

From a human viewpoint we would probably argue that most of these changes have been good. After all, nearly everyone lives in environments of exotic species and we obtain our living from them. But there is a big price to pay for this – in fact it has caused two of our largest conservation problems which are studied later:

1 **Unwanted invaders** (pages 18–19)
2 **Loss of native species** (pages 20–21)

This farmland has almost no native plants.

Activity: Conflict over the logging of native forests

In the late 1990s a very bitter conflict took place between groups who felt that any logging of native forests was wrong and Timberlands, a government owned company which was involved in selective logging of native forests on public land on the West Coast. In 2000 a new government stopped all logging in the native forests.

1 In your groups:

a Discuss the arguments put forward by each group. For each argument decide how it supports that group's interests.

b Decide which view you would take and why.

2 Explain why the following were done:

a Timberlands employed a publicity company.

b Some protestors spent time perched in trees.

c Both groups spent time and money 'lobbying' the government. (You may need to find out what 'lobbying' is.)

3 Sustainable milling – what does this mean? Why is it important?

4 Many West Coast people were unhappy with the new government's decision. Explain why.

Timberlands' arguments:

1 Selective logging using helicopters does not destroy the forest and is sustainable.
2 Valuable native timber is produced
3 Logging provides jobs – scarce on the West Coast.
4 Most of the forests will not be milled.

Protestors' arguments:

1 What little that is left of these ancient forests is our children's future.
2 The forests are the habitat for endangered native plants and animals.
3 The logging methods being used seriously damage the forest.
4 There are plantation forests to supply our timber needs.
5 Ecotourism in the forests is a long-term alternative industry.

11 The invaders

The list of plants and animals from overseas (exotic species) that have arrived in New Zealand because of humans is very large, and their effects are enormous. Some, such as sheep, cattle, and farm and garden crops we could hardly live without. Others have had very harmful effects on our environments. They:

1 **Eat the forests:** Plant-eating animals have found our native plants very tasty, and with no enemies they have eaten their way through much of our forests.
2 **Kill native species:** Carnivorous animals have found our native birds and reptiles, which evolved without enemies, easy to catch and kill.
3 **Replace native species:** Many exotic plants are fast growing with no natural enemies in New Zealand and they often choke out native plant communities.

Possum – the number one problem

The Australian brush-tailed possum looks really cute and harmless, yet it is by far our most serious problem species. As early as the 1860s the possum was brought into New Zealand several times to start a fur industry. It took quite a long time for it to spread, but it is now common in most of the country. In New Zealand the possum is much larger and grows and reproduces faster than in Australia.

Why are they a problem?

1 There are over 70 million possums in New Zealand. They munch through 20,000 tonnes of forest each night.
2 Possums found the New Zealand leaves of native bush plants very good to eat (remember the bush had developed without browsing animals). In many forests most of the more tasty trees and shrubs have disappeared.
3 The possum is a carrier of bovine tuberculosis (TB), a serious disease of cattle.
4 Possums are also a threat to native birds. They raid their nests.

What can be done about them?

1 **Trapping and shooting:** This has been carried out on a large scale, but it is expensive and has not succeeded in keeping the possum in check.
2 **Poison:** This is the most effective method of control. But the 1080 poison used is harmful to humans and other animals.
3 **Biological controls:** Scientists are working both on finding organisms that will be harmful to the possum, and a chemical that can affect its reproduction. But these solutions also have problems. For example, in Australia the possum is protected. If a biological control agent reached there it could wipe them out.

New Zealand's tally of exotic arrivals

1 Mammals:
Thirty-four species have been introduced.

- Sheep, cattle and other farm animals are very common and very important to us.
- Herbivores (plant eaters) such as possum, rabbit, goat, deer, and rats (scavengers) cause terrible problems.
- Carnivores (meat eaters) such as ferret, stoat, feral (wild) cats and sometimes dogs are serious problems to native species.

2 Birds:
Over 30 species of mostly small birds (sparrows, thrush, blackbird, etc.) were introduced in the 19th century and are now the most common birds in most of New Zealand. Several such as mynah and magpie have become problems.

3 Insects:
Nearly 1000 different kinds of exotic insect have been found in New Zealand. Many are a serious problem, such as wasps, white butterfly, green vegetable bug, and some flies, mosquitos, cockroaches and ants.

4 Plants:
There are about 2000 alien plant species in New Zealand. Many were brought in, others such as most weeds, came accidentally. Perhaps as many as 200 cause harm – especially gorse, broom, ragwort, hawkweed, old man's beard, privet and wild ginger.

Activity: Getting to grips with the possum

1 The possum problem in New Zealand
In Australia the possum is a protected animal.

a List three ways that the possum in New Zealand is different from in Australia.
b List at least three reasons for these differences in the two countries.

2 Measuring the problem
Each possum eats 300 g of forest each night.

- Weigh out 300 g of leaves.
- Find your weight. Calculate how many possums would eat your weight each day.
- Work out how many students would weigh about 20,000 tonnes (the amount eaten by possums each day).

Plant pests: Broom

In spring and early summer the hills near Hanmer in North Canterbury are covered with a dense carpet of yellow, stretching for many kilometres. This is broom. It has escaped from around the homes of the early settlers, and because it grows so fast and strongly it has taken over the once tussock-covered hills. It is so thick that the hills are now useless for pasture, and it is too costly to get rid of. Broom and gorse are just two of the many plants that were brought to New Zealand and have become pests in some places.

Broom.

Animal pests: Stoat

This stoat is a victim of a car. It is a sight that is becoming more common on roads as the stoat population increases. Stoats (along with ferrets and weasels) were brought to New Zealand by early settlers in an attempt to solve another problem – rabbits. Rabbits had been brought to the South Island because the settlers thought it would be nice to have some around for sport and to remind them of England. The rabbit, with no enemies in New Zealand, quickly spread and became a serious pest. So the stoat which feeds on rabbits in England, was brought to New Zealand. Unfortunately the stoats found our native birds much easier to catch. They have spread and are now one of the most serious killers of birds.

Dead stoat.

Insect pests: German Wasp

We all know what a pest this wasp can be. It arrived accidentally in Hamilton in 1944 when nests were found in cases of aeroplane parts from England. At first it spread gradually, but in recent years it has become a serious pest. It has invaded the beech forests where it feeds on honey dew, and in summer it can be so common that it is not safe to be in the forest. The nests are underground or in trees and can contain many thousands of wasps.

Activities: Getting to know the invaders

1 Understanding the problems

Write down why each of these sorts of animal or plant were not problems in their homeland, but quickly spread and became pests in New Zealand:

a Plant eaters like rabbit and deer.
b Small meat eaters like stoats
c Plants like gorse and broom.

2 Researching a problem invader

Choose a plant or animal from overseas that has become a problem in your local area. Do research on it and write a page report on it, using the following headings:

- Description of the species (drawing if you wish).
- Its habitat (where it lives in the district).
- Why it is a pest.
- Where it came from.
- How it got to New Zealand.
- What can be done to get it under control.

Around Auckland common plant pests you could choose are tree privet and wild ginger plant. Gorse is common around Dunedin and Wellington. Each local area will have its own problem plants and animals. Check the box on page 18 – there are many others.

3 Getting the message across

Imagine you have been asked to design a publicity campaign to let people know about a plant or animal that is becoming a pest in your area. Choose a suitable plant or animal. You can:
either design a poster
or outline a TV ad (sketch or make notes of what you would show).

4 Thinking about attitude changes over time

Either write down or discuss in your groups, the reasons why people did the following, and why our view might now be different:

a The early settlers brought out hundreds of English birds (for example, sparrows, thrush, blackbird).
b Rabbits were brought from England.
c Stoats and ferrets were brought from England.
d Gorse was brought from England and planted as hedges.
e Some people still secretly release deer in forests that were free of them (such as in Coromandel).

Exotic Pacific Oysters cover the rocks on this Auckland beach.

12 Species in danger

The list of New Zealand's endangered species reads like a horror story. New Zealand's native birds in particular are suffering badly from having evolved for 80 million years in isolation from predators. The kiwi is a good example.

The kiwi – is our national bird going to disappear?

Our national bird is in serious trouble. There were once many millions of them living in our forests, but now there are only about 50,000 left on the mainland of New Zealand. Most worrying is that recent studies show that it is declining fast – at a rate of 6 per cent every year, and that 95 per cent of all chicks hatched do not survive for more than six months.

Why is this happening?

A study in Northland using radio transmitters to monitor birds showed:

1 Most chicks are being killed by stoats – the worst problem.
2 Possums are interfering in the nesting period.
3 Adults are being killed by dogs and ferrets. One dog alone is reported as having killed 500 kiwi in one six-week period in a Northland forest!

So, although the kiwi is smart and aggressive, it is not designed to deal with the new killers who use their sense of smell to hunt on the ground.

What can be done to save the kiwi?

A Kiwi Recovery Programme has been launched to help the kiwi. It involves:

1 **Research:** This is being carried out to determine the numbers of kiwi and how they live.
2 **Operation Nest Egg:** Eggs are taken from the nests and hatched and reared in a protected place (for example, Auckland Zoo). When the chicks are six months old and able to defend themselves from cats and stoats they are released back into the forests.
3 **Stoat control:** $6.6 million has been provided for research into the best ways to control these very smart animals.
4 **Education:** DoC experts visit farming areas explaining to local people what the problem is and how important it is to keep dogs tied up and to trap predators.

Can the kiwi be saved? So far there are encouraging signs that this programme is starting to work. But is it too late?

New Zealand's tally of extinct and endangered species

- **Natural habitats:** Only about 35 per cent of New Zealand is now covered with native communities. So native species have lost most of their natural habitats.
- **Gone forever:** Over 30 per cent of our land and freshwater native birds are now extinct (43 species) and 18 per cent of our seabirds. About 25 other kinds of animals and 11 kinds of plants have gone.
- **In danger of extinction:** *About 800 species are in danger of extinction.* These include many of our smaller plants, some small animals, frogs, lizards and most of our birds.
- **Birds in danger:** They include three-quarters of all our species of native land birds. The most endangered include kakapo (only 86 left), takahe (200), black robin, black stilt, kokako, some kiwi species, fairy tern, dotterel, teal duck.

One hundred years ago there were about 5 million North Island Brown Kiwis. Now (2000) there are only about 30,000, and there are nearly 6 per cent fewer every year.

Activity: Becoming informed about the problem

1 The kiwi's adaptations
The following are adaptations of the kiwi that helped it to survive in the past. For each explain if it:
A helps or
B hinders its survival at present. Give reasons.

a It is tough and aggressive.
b It can not fly.
c It is a smelly bird.
d It has a long beak, keen hearing and good sense of smell.
e It lives in holes on the ground.
f It breeds very slowly.
g Its powerful legs kick strongly.

2 The kiwi's problems

a From the notes above list three problems the kiwi is facing.
b List four things being done to help save the kiwi. For each write a sentence explaining how it may help.

3 Other species in danger
Choose one of the seriously endangered species and produce an information sheet about it and its problems (poster or project). Make sure you explain:
- Details of the species – where and how it lives.
- How endangered it is and why.
- What is being done to save it.

Use your library, web and books for information.

Species in danger overseas

The loss of biodiversity is a big problem all over the world. It has been calculated that right now Earth has about 30,000 species in danger of extinction. It is possible that up to 100 kinds of living things are actually becoming extinct every day as a result of humans. Many extinctions occur overseas for the same reasons as in New Zealand – habitats destroyed, introduced predators or other species that take over. But there are other problems also:

1 **Hunting** for food or sport until none are left – This used to be a big problem and has resulted in the loss of many species, especially African game animals. This is much less of a problem now as people around the world are beginning to appreciate biodiversity.

2 **Poaching** for parts that can be sold for high prices – In many countries, especially in Asia, parts of some animals are highly valued as medicines (e.g. rhino horns), or for their skins (e.g. tigers). The tusks of elephants are valued for their ivory. Although killing elephants for their tusks is banned, poachers kill hundreds each week. They receive about $60 per tusk from smugglers who send them to Asian countries where they may be worth $1500.

Biodiversity – what is the problem?

Biodiversity is the range of all the different kinds of living things. Why should we worry if the odd species dies out and biodiversity becomes reduced? Most plants and animals that become extinct do not seem to be very important or useful to us anyway. They are always replaced by species that seem hardier and better able to cope. It is easy to argue this way, but there are some other viewpoints. Think about these:

Economic: Species that have not yet been studied - or even discovered - are the only source of genetic diversity. This means new varieties and new chemicals that may be useful to us in the future.

Scientific: We need to study rare species in order to find out more about how the natural world functions.

Moral: Humans now have the power to destroy the natural world and its wild life which we have a duty to look after for future generations.

Beauty: Living things are interesting and beautiful and once they are destroyed they can not be replaced.

There are only 5000 tigers left in India, 300 of these are poached each year. At this rate when will they become extinct (assuming that apart from poaching, birth and death rates are the same)?

Activities: Questions about saving biodiversity

1 Analysing biodiversity

Discuss in your group, or as a class, what you think about the four arguments put forward (in the biodiversity box) about why endangered species should be saved.

2 The ivory poaching problem

A The killing of elephants for their tusks has been banned. Yet poaching still goes on. Explain the views of these people involved:

a The poachers.

b The park rangers who protect the elephants.

c The smugglers who get the ivory to Asia.

d The merchants who sell carved ivory.

e The tourists and others who buy the carved ivory.

B How can this trade be stopped and what are the problems?

3 Saving endangered species in New Zealand

The Department of Conservation has become a world expert in working out ways to save endangered species. Some of these methods are explained in the notes on the kiwi. Another is putting endangered species on offshore islands.

a Find out where these sanctuary islands are and mark them on a map of New Zealand – Little Barrier, Tiritiri-matangi, Kapiti, Stephens, Mana, Codfish, Resolution.

b Explain why offshore islands would be useful as sanctuaries.

c Explain why these islands need to be regularly checked for predators such as rats.

d It costs many millions of dollars to do this work. The kakapo saving project alone is costing several million every year. In class or in your groups debate the following:

- Is it worth the cost to save a few endangered birds?
- Who should pay – the government, sponsors from business, donations?

A little South Island Robin. Is it worth saving?

13 BIOSECURITY – PREVENTING INVADERS

Protecting the long borders of New Zealand from unwanted invaders is a huge task. In 1993 a Biosecurity Act was passed, and the Ministry of Forest & Fisheries set up a quarantine service to prevent the entry of any pests or diseases that may harm our environment, our agriculture or our health. The MAF quarantine service looks for:

- **Live animals** – cats, dogs, reptiles, birds.
- **Food** – fresh fruit, meat, vegetables, dairy products, honey.
- **Plant matter** – live or dried, cuttings, seeds.
- **Animal products** – feathers, bone, skins, shells, wood and items made from them.
- **Straw items** – baskets and mats made of plant materials.

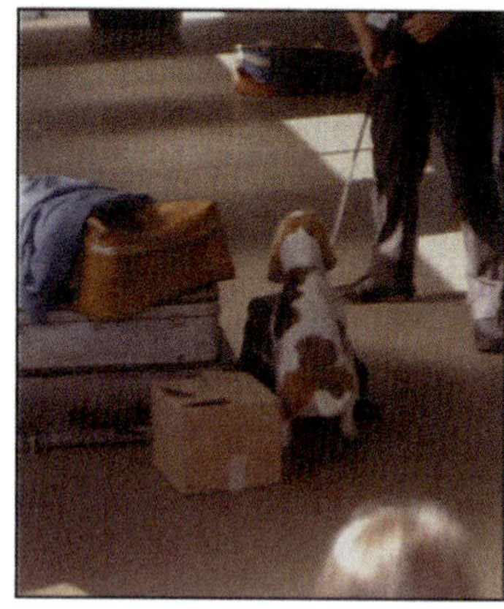

In one month the beagle brigade searched 74,000 passengers and their luggage and made 495 seizures. Five-year-old Benny searches over 1000 passengers per day. Among his discoveries were six tortoises, a cat shut in a bag by mistake and a dog in a handbag!

The Beagle Brigade

These dogs are the best known part of the quarantine checking system. At the airports small, friendly beagle dogs use their fine sense of smell to check people and bags that contain illegal items, especially foods. At the overseas mail-sorting centre a team of mongrel dogs do a similar job on the 230,000 items of mail that arrive each month. They are specially trained to retrieve mail with plant or animal matter in it. The officers also use X-ray machines on baggage and computer profiles, details of background, where from, and so on, to identify high-risk passengers. They become very experienced at deciding who to check. Each year about 80,000 seizures are made.

The escapers

A few unwanted invaders do slip through the net. They are mostly insects and they usually arrive on cargo coming through ports – in containers, on cars, and so on. When a new invader is found, MAF teams very quickly make an assessment of the problem and if possible carry out spraying programmes to kill the pest. In 1996 the tussock moth was found in Auckland. Because this was likely to be a very serious pest $6.6 million was spent to get rid of it. Part of the city was sprayed several times from a low-flying aeroplane. Two years later an Australian mosquito that carries some unpleasant diseases was found breeding in wetlands around the port of Napier. A helicopter spraying programme was carried out to destroy the mosquitos. Unfortunately, in spite of this, two years later the mosquito has been found in further sites in Gisborne and around the Kaipara Harbour near Auckland. Each year more possible problem invaders like this arrive in New Zealand. Foot and mouth disease has never affected New Zealand's farm animals. The devastating 2001 foot and mouth outbreak in England is an example of why it is so important to protect our borders.

Spraying for the tussock moth over Auckland city in 1996.

Activities: Studying our defences

1 What can be allowed in?

The MAF officers at the airport may:

A Allow an item through.
B Get it fumigated and then let in.
C Have it destroyed because it may carry pests or disease.
D Have it destroyed because it helps species to become extinct (e.g. ivory).

The following items have been brought in by passengers on a plane from Singapore. Decide what would be done to each item – and explain why:

a bananas b wooden face mask
c bunch of flowers
d carved ivory beads
e tin of baby food
f live parakeet g jar of honey
h woven straw mat
i unopened packet of biscuits
j basket made of dried grass

2 Why is biosecurity important?

Answer the following:

In what way does being an isolated island make it

a easier b harder

to run a good biosecurity system?

c New Zealand has one of the strictest biosecurity systems in the world. Explain why.

d Explain why the following are problems for the MAF quarantine service:
- cargo containers
- the small boats that arrive round the coast from overseas
- arrivals from countries with a similar climate to New Zealand

e Most new invaders first tend to appear near ports or airports. Why?

3 The biosecurity system

a Explain why each of the five types of item listed at the top of the page could be harmful.

b Explain why each of the following are used:
- dogs
- X-ray machines
- computer profiling of passengers.

4 Publicising biosecurity

Design a poster explaining to people *either* what can and can not be brought into the country *or* why we need a MAF quarantine.

Reference: www.quarantine.govt.nz

Activities: Local issues

Protecting natural environments is not just a problem in other parts of the world (for example, the rainforests) or in other places in New Zealand. There are many issues in your own district – and even around your own home or school. Here are a couple of ways you and your class can be involved:

A Dealing with a local issue

1 Choose an issue
In your local community there will be many local problems about either the protection of an endangered species or dealing with an unwanted predator or weed that is spreading. (Examples include Hector's Dolphin around Banks Peninsula, rabbits in central Otago, privet in the bush in Auckland city, and many others.) The local Department of Conservation or a local conservation group will be able to help.

2 Research the issue
Find out all you can about the issue – where, why and how serious it is, what is being done, and if there are different viewpoints about solving it. To get the answers you will need to check local newspapers, talk to local experts, perhaps visit the places. This will need careful planning.

3 Publicise the issue.

4 Take action
The action you can take will need discussion. You could:
- Raise money to help.
- Have a field trip – visit the problem area.
- Have a working bee to remove weeds.

B Improve the school environment

(This could also apply to your own home or to a local waste area.)

1 Evaluate the state of the school environment
Check out and discuss the school environment. Is it well cared for? Are there parts that lack trees, shrubs or gardens? Has it lots of native plants – especially those that have berries or nectar to attract birds? Can it be improved?

2 Plan your project
Get advice (and permission) about planting from experts – the groundsman, principal, board of trustees, interested teachers. Design and plan carefully what you can do.

3 Grow plants from seed
If you can obtain seeds of suitable plants that have grown locally you may be able to produce your own seedlings. (You may prefer to leave this step out if you can obtain seedling plants.) You will need to get advice on suitable native shrubs for your area and on how to plant and care for seedlings. (The Forest and Bird Society has a work sheet on growing native plants. www.forest-bird.org.nz)

4 Planting
When you have suitable plants large enough you will be able to plant them out at the right time of the year. You will need advice on how to prepare the ground and to plant the shrubs.

5 Long-term care
It is important that your plants continue to be cared for. They will need to be checked regularly and when necessary weeded, fertilised and watered.

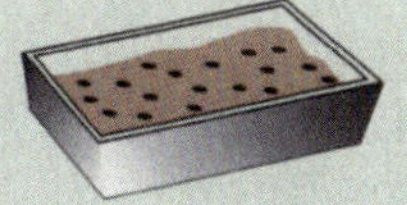
Put seeds on damp soil in trays

Cover seeds with 1 cm of fine soil

Keep damp and cover until seedlings appear

What can you do about protecting our natural environment?

1 Get to know your natural environment.
- Visit National Parks and other wild places.
- Visit biodiversity organisations – e.g. local zoo or wildlife park.
- Visit biosecurity protectors – e.g. MAF quarantine service at airport.

2 Get involved with local conservation groups.

3 Take part in school and community tree planting and weeding projects.

4 At home, plant trees and shrubs in your garden – especially natives that produce nectar or berries.

5 Grow your own native plants from seed.

Resources under pressure

14 Water, the vital resource

Of all the resources humans need from the Earth, water is the most vital. Yet it is one that we have never worried about because there seemed to be an endless supply. Unfortunately, fresh, drinkable water has suddenly become one of our greatest concerns for the new millennium.

- Humans already use 54 per cent of the Earth's fresh water.
- *Nearly 40 per cent of the world's people (2.3 billion) already face regular water shortages.* (Imagine 12 students in a class of 30 not getting enough water.)

When you remember that another 3 billion people will be on Earth and needing water in 50 years time, you can see how serious the fresh water problem is.

What have we done to our endless water supply?

Although 70 per cent of the Earth is covered with water, 97.5 per cent is salty, and much of the fresh water is frozen, leaving only 1 per cent for our use. Of course, the water cycle means that water is always being evaporated and recycled as rain. But humans have interfered with this in several ways. Look at the diagram:

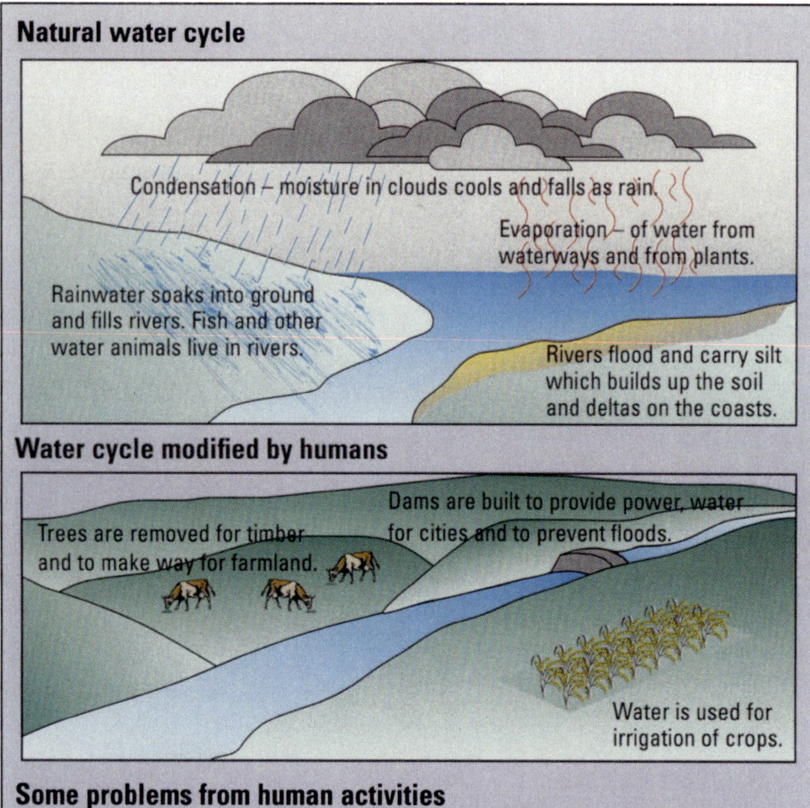

Some problems from human activities

1. **Reduced rain** – droughts more common and wells may dry up.
2. **Silt** – not dropped downstream. Deltas shrink. Silt builds up in dams.
3. **River flow reduced** and lakes and wetlands shrink.
4. **Salinisation** – soils become salty and useless in dry areas if soaked in water too much.
5. **River fish reduced**. Fish such as salmon, which swim upstream to lay eggs, may be severely reduced in numbers.

Worst drought in 100 years saps life out of rural India

JODHPUR – In many of the villages of western Rajasthan the rivers and ponds have dried up, the rainwater tanks are empty, the water table in the wells plummets.

Hundreds of years ago, the villagers prepared against droughts by digging wells lined with stones in the beds of the seasonal ponds. At first they guarded these bheri or wells fiercely, but when the government put water pipelines into the desert they forgot about them.

Because the pipes have not been looked after and sometimes have no water, the village women are rediscovering the bheri. But even they are almost dry.

The bheri is a round pit three metres deep. The big spherical pots of the women who have reached the front of the queue fit at the bottom. Water oozes very slowly down the sides. Wait half an hour and a shallow, muddy puddle has formed in the bottom. Pour it into the pot with a scoop, clamber out, and squat in the shade to wait for half an hour or more.

Not surprisingly many women give up in disgust. The water is dirty anyway. The only other way to get drinking water is to buy it from sharp local businessmen. They truck it in and sell it for high prices.

From an article in *NZ Herald* April 2000

Dams – good or bad?

In the last 40 years humans have tried to control the world's water resources by building huge dams across most of the world's rivers and creating large reservoirs. In fact, man-made lakes now cover an area more than one and a half times the area of New Zealand! Unfortunately we are now beginning to realise that often the environmental problems caused by huge dams are greater than their benefits. In fact, in some places, particularly in the United States of America, dams are actually being removed.

China's Three Gorges Dam project

China is at present working on the Three Gorges Dam on the Yangtze River. If this dam is completed it will be the largest in the world. Its scale is mind boggling.

- The height of the Three Gorges dam will be 175 metres – half the height of Auckland's sky tower!
- Its width will be over 1½ km. (It would take over 15 minutes to walk across it!)
- It will create a lake 600 km long. (Further than Auckland to Wellington!)
- It will cost over $25 billion and take 20 years to complete. (It started in 1994.)

When it is complete, the Three Gorges Dam will change the traditional way of life for millions of people in China.

The upside of the project

The dam is being built for two main reasons:

1 To greatly reduce China's dependence on expensive and polluting coal power by suppling large amounts of hydro-electric power.
2 To control flooding in the mighty Yangtze River – the third largest in the world. The rich Yangtze plain regularly suffers from such severe flooding that about 300,000 people have been killed by floods in the last 100 years.

The downside

All over the world concerns have been expressed that the harmful effects will be too great. Here are some of the downsides:

1 Several large cities will be flooded and over 1.2 million people are being forced to move and resettle in new places.
2 Many valuable historical archaeological sites will be covered.
3 Huge amounts of industrial waste and sewage will pour into the new lake and will not be flushed away as in a river.
4 Silt now carried by the fast-flowing river will no longer be carried downstream to enrich the soil on the floodplain.
5 The silt will build up behind the dam and block the spillways.
6 The habitat of many species of fish and dolphins will be threatened.

Activities: Water problems

1 Explaining the water cycle diagrams

a Why is the natural water cycle diagram referred to as a cycle?
b What causes water to evaporate from the plants?
c List three advantages to humans of building dams.
d How have forest removal and irrigation helped to provide more food?
e For each of the five problems listed, write a sentence explaining what human activity has caused this problem.

2 Surviving India's water shortage

Look at the newspaper cutting.

a The first villagers guarded the bheri carefully. Why?
b Why did the villagers forget the bheri?
c The villagers who cannot afford to buy water are fleeing to the cities. Why?
d Who are the people making a profit from the drought?

3 Weighing up the Three Gorges project

Should this project be allowed to be completed? In your groups discuss each of the upsides and downsides listed.

a For each decide if it is a strong argument.
b Decide if you feel the project should go ahead.
c How you would feel if you were told that you were being moved from your home and job and forced to live in a new town where you had to start a new life?
d The Yangtze river is called 'China's sorrow'. Why?

4 Dam problems

For each of these problems with water explain why this is happening:

a Since the rivers were dammed in western Canada, most of the salmon fishery has gone.
b The swamps of the Everglades in Florida are drying up since rivers were tapped for irrigation.
c A dredging programme will be needed in Lake Dunstan in central Otago now that the Clyde Dam is operating.
d It has been estimated that more land is being lost for farming by salinisation than is being gained by new irrigation schemes

15 Resources from the sea

Humans have always fished the oceans of the world as if they are an endless resource. But as the population of the world has grown, so has the demand for fish and the seas have been fished more and more heavily. Fishing methods became more efficient until quite suddenly, in the 1970s, what happened to the Peruvian fishery (see graph), happened to several other important fishing areas. In Newfoundland 35,000 people lost their jobs when the cod fishery collapsed. No longer was the old saying that 'there are plenty more fish in the sea' true.

Over 60 per cent of the world's fish stocks are now under pressure. Large trawlers, using the latest technology, are searching over the world to find new sources of fish. Unfortunately, there are few places left where the fish are not being exploited.

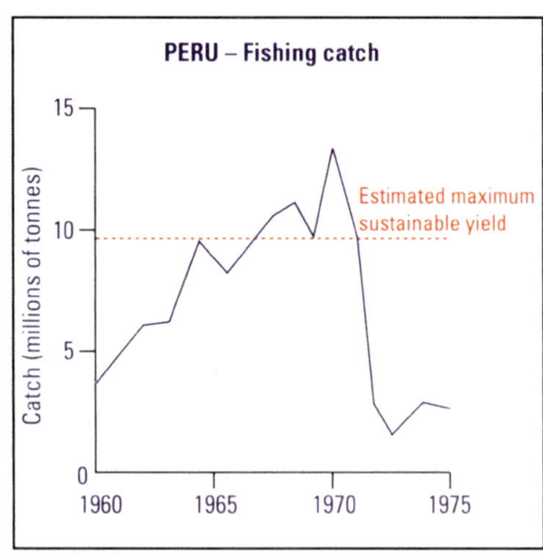

Before 1970 the Peruvian anchovy fishery caught 10 per cent of the world fish catch. As fishing technology improved more and more fish were caught. Then in 1970–71 the catch became too large. The breeding population was destroyed and the fish catch crashed. It took over 20 years before it began to recover.

The need for sustainable fishing

The problem now is that there are too many fishers chasing too few fish. The number of people fishing commercially has doubled since 1970. No longer can the world afford to let fishing be a 'free for all'. The world's fisheries must be managed sustainably. This means that they must not be fished more heavily than the fish stocks can be replaced. This is easier said than done. Before it can happen there are two big problems that need to be overcome. They are to do with:

A Fishing methods

Improved technology means that fishing by the big commercial fishing nations is so efficient that these things happen:

- **overfishing** – areas are quickly fished out and there are not enough fish left to breed and rebuild the numbers.
- **waste** – it is wasteful because everything is usually caught – if it is wanted or not. About 20 million tonnes of unwanted fish (by-catch) are thrown away each year. This is nearly 25 per cent of the world catch wasted.

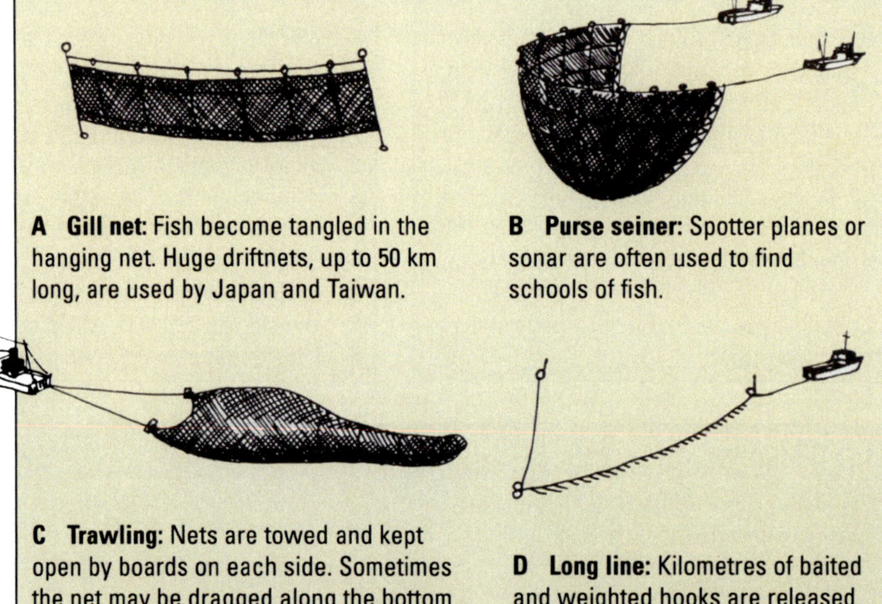

A Gill net: Fish become tangled in the hanging net. Huge driftnets, up to 50 km long, are used by Japan and Taiwan.

B Purse seiner: Spotter planes or sonar are often used to find schools of fish.

C Trawling: Nets are towed and kept open by boards on each side. Sometimes the net may be dragged along the bottom like a dredge.

D Long line: Kilometres of baited and weighted hooks are released on a line.

B Fishing rules

International agreements have been made to decide on:

- **quotas** – how many fish are caught and by which country.
- **exclusion zones** – areas set up where fishing is banned to let numbers recover.

But fishing nations cannot agree over these rules. Norway and Japan, for example, refuse to agree to rules not to catch endangered whales.

New Zealand has had major problems with overseas fishing vessels exploiting fish in New Zealand waters. In 1990 the Japanese and Taiwanese had 20 vessels using huge driftnets to hunt for tuna in the South Pacific. As a result of protests from Greenpeace and others, driftnet fishing was banned in the South Pacific.

Activity: Ways of catching fish

Fishing methods:

A Sketch each of the fishing methods shown above. Under each, note which type of fish they would be used for:

1. school fish such as kawhai
2. open sea fish such as tuna (2 methods)
3. bottom fish such as snapper
4. shellfish such as scallops.

B Which methods might cause these environmental problems:

1. Causes waste by catching marine mammals and all kinds of fish apart from the kind wanted.
2. Catches diving seabirds such as albatross in the hooks.
3. Destroys the seabed habitat.
4. Wasteful because many trapped fish die and may rot.

New Zealand's fishing industry

New Zealand has a very large area of territorial waters (200 miles from land) and quite a large fishing industry.

Unfortunately, New Zealand's record of managing its harvest from the sea in a sustainable way has been poor. In the past we have **exploited** rather than **managed** our resources.

First the early sealers destroyed a huge population of seals. Then the whalers did the same to the whales. The whitebait and scallop fisheries have declined. The Bluff oyster quota has been greatly reduced and for several years there has been no toheroa gathering season. The crayfish industry had a boom in the 1960s when crayfish were found around the Chatham Islands. This collapsed and has remained at a low level. The same has happened to fish such as snapper.

Although New Zealand now has quotas for catching most sea life, these are not always agreed to by the fishing industry and often may still not be low enough to sustain the fishery.

The story of the orange roughy

The orange roughy has been one of our main commercial fish for the last few years. Is it the next fishing disaster? In 1981 large numbers of orange roughy were discovered living in deep water near the Chatham Islands. Within a few years a large commercial catch was built up and became a valuable export, mainly to the United States (over $1 million in 1996). By 1988 the catch peaked and since then it has dropped rapidly.

This happened before scientists could study the fish and decide what was the MSY (maximum catch that would allow the numbers of fish to be sustained). Because the orange roughy lives as much as 1 km beneath the surface it is very hard to study.

By 1992 MAF scientists had discovered that the orange roughy is a very slow-growing fish, living for over 100 years. It may not reach breeding age for over 25 years. They felt that, by 1992, perhaps 80 per cent had already been fished out. The quota limit has been reduced, against the wishes of many in the fishing industry, but the catch continues to decline. In 2000 some areas are reported to be down to 3 per cent left!

Have we destroyed this resource too?

Orange roughy annual catch

Year	Tonnes caught	Quota
1988	62,000	
1990	48,312	47,537
1992	37,012	38,117
1994	29,688	35,490
1996	20,700	21,330
1998	16,664	21,330
2000	18,546	20,365
2002	14,379	15,421

Activities: Studying the fisheries

1 Definitions

a Write down the meanings of these terms (see pages 26 and 27): sustainability, quota, maximum sustainable yield (MSY).

b For each explain why it is important to the fishing industry.

2 The Peruvian fishery collapse

a What happened to the fishery in 1970?

b Give two reasons explaining this.

c How long did it take before the fish stocks began to recover?

3 Conflict about quotas

In your groups discuss the following two arguments about quotas. For each decide:

a If it is a good argument.

b What your reply would be.

Commercial fishermen say: 'We want the quotas to be kept high because our jobs depend on catching and selling fish.'

Fishery scientists say: 'We want the quotas kept below the sustainable level otherwise there will be not enough fish in the future.'

4 Data plotting and analysing

The following is the estimated biomass (total weight) of snapper in Hauraki Gulf/Bay of Plenty areas over the past 140 years:

Year	tonnes of fish
1860	265,000
1880	255,000
1900	240,000
1920	195,000
1940	70,000
1960	50,000
1980	30,000
2000	30,000

a Plot this data on a graph. (Put years on the horizontal axis.)

b What happened to the snapper stocks between 1910 and 1940?

c What is the present situation?

d It is estimated that the MSY is about 65,000 tonnes. If this is so, what is likely to happen in the next few years?

e Snapper is a popular catch for recreational fishers. Should these people have a quota? Why?

5 Analysing the orange roughy

a Draw a graph to plot the annual catch of the orange roughy. (Put years on the horizontal axis.) Also plot the quota.

b How does this graph compare with what happened to the Peruvian anchovy fishery?

c Is it fair to say that we do not appear to be managing this fishery well? Why?

d Why is it difficult for scientists to get reliable estimates of the fish numbers?

e Why is the fact that the fish take 25 years to mature important?

6 Aquaculture

This is fish farming – either in ponds, rice paddies or in cages in the sea or river. It now accounts for over 16 per cent of all fish caught, and is growing rapidly. Explain why.

16 Supplying our energy needs

From the time the early humans began to use fire for warmth we have used different kinds of energy to make our lives better. Finding wood to burn is still an important job in the poorest countries. Modern developed countries are built around the use of large amounts of energy. We want it to make our homes comfortable, to run our industries and to power our transport. As the population of the world continues to grow and more people can afford a higher energy-using lifestyle, energy demands will keep increasing. (Electricity demand alone is expected to increase by 2.5 per cent per year until 2020.)

Cooling towers for a nuclear power plant.

Fossil fuels

At present over 80 per cent of all the energy needs of the world are met by burning the fossil fuels – **coal**, **oil** and **natural gas**. These are formed from the decay and compression of plant and animal matter in the earth over millions of years. But burning of fossil fuels cause two big problems:

1 **Fossil fuels are non-renewable resources.**

Once they are used up there are no more. At present this is not too big a problem because there are still many large reserves of coal and oil not yet exploited, but new sources are becoming harder to find.

2 **Fossil fuels pollute the air as they burn.**

They release fumes containing dirty and harmful substances such as sulphur. They also release large amounts of carbon dioxide into the air. This is a special worry because evidence increasingly suggests that this is contributing to global warming (see page 42).

Wind farm.

How can we solve the energy problems?

The estimates of future energy use suggest that the people of the world have not yet seriously faced up to these energy problems. To solve them there are two things we must do in the future:

A Use renewable energy sources:

These are energy sources that do not run out. It is essential that they be used much more. It is predicted that the amount of renewable energy use will increase by around 55 per cent by 2020, but this is simply keeping pace with increased fuel demands. We will need to do better.

Nuclear energy has not fulfilled its promise because it is too expensive and dangerous, and **geothermal** energy sources are rare. Even **hydro-electricity** is now seen as having downsides related to its effects on rivers. (See last unit.)

The greatest hope for the future of renewable energy use is probably with solar and wind power -both renewable and pollution free:

1 **Solar energy:** This is trapped directly from the sun, but it is still expensive. A great deal of research is being done on designing solar cells that trap the sunlight more efficiently than at present.

2 **Wind:** The energy from wind is caused by the sun's heating of the Earth and the air above it. New designs of wind-driven turbines are rapidly becoming more efficient, and at present they look the most promising of the renewable energy sources. In fact, the amount of wind generated power increased by nearly 40 per cent in 1999 alone. New Zealand has two wind farms – in the Wairarapa and near Palmerston North.

B Increase efficiency in using energy:

Our use of power is at present very wasteful. There are two ways we can use it much more efficiently:

1 **Use energy less wastefully:** In industry and in our homes we can do such things as improve insulation in buildings, use energy-efficient light bulbs, car pool, and much more to make sure that we do not use so much energy (see page 31).

2 **Design more efficient power generators:** Some most promising research is being done on improving the efficiency of cars – developing experimental cars that run on fuel cells (based on the use of hydrogen) rather than petrol burning motors.

Solar energy farm.

Sources of energy

Sun – source of heat and light

Photosynthesis converts solar energy into plant and animal matter

A Geothermal

Renewable energy from hot rocks underground changing water into steam. Not common.

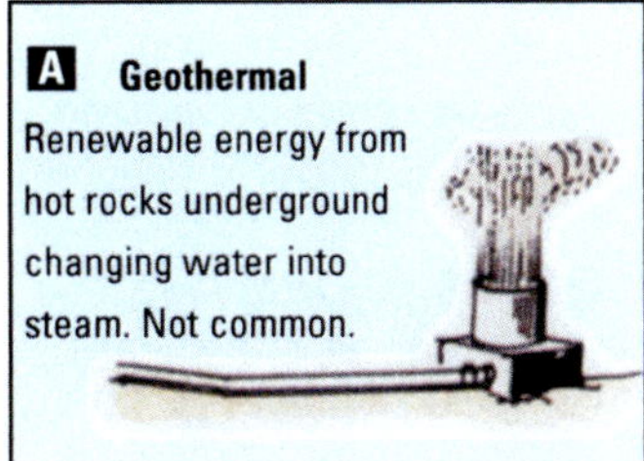

B Nuclear

Energy from splitting atoms. It is costly and produces radioactive wastes.

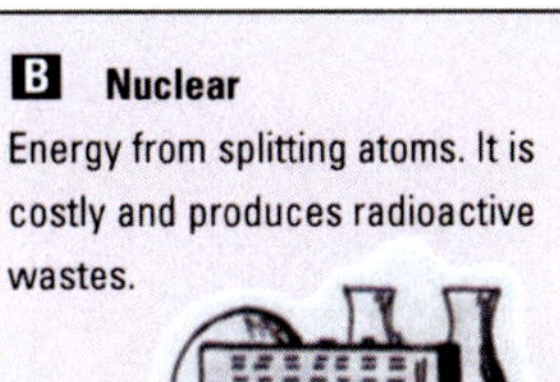

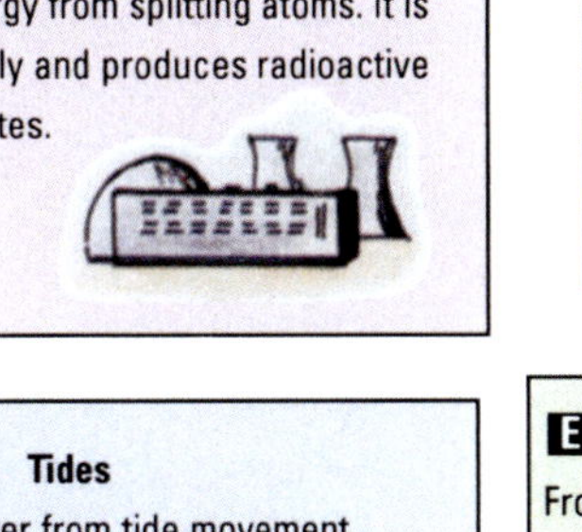

C Tides

Water from tide movement trapped in some estuaries. Tide caused by gravitational pull of moon. Not widely used.

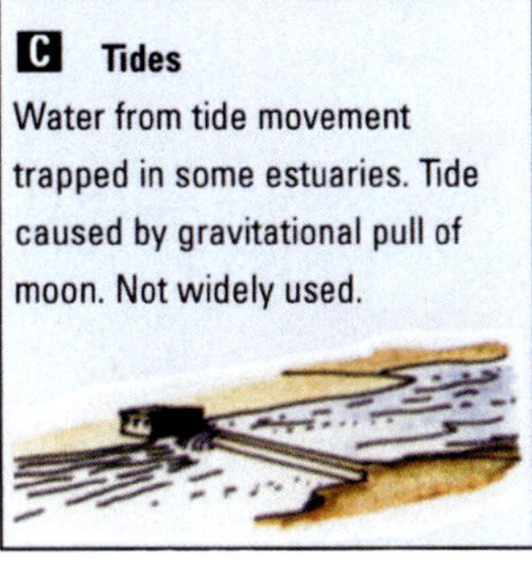

D Solar

Sun's energy trapped by solar panels. Needs sunny places. Still quite expensive.

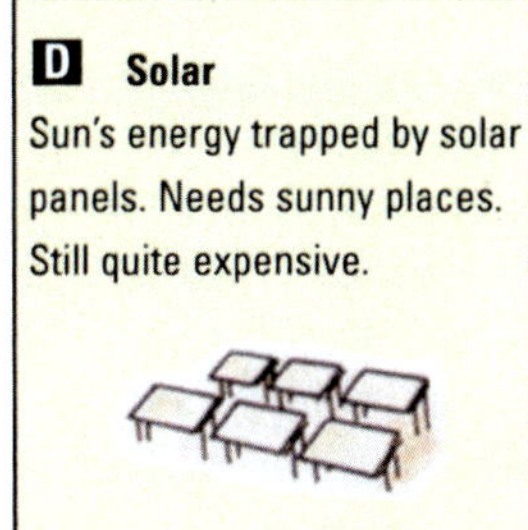

E Hydro-electricity

From force of falling water. Clean but dams affect water flow and cause silting.

F Wind

Movement of air caused by sun heating and altering air pressure. Turbines need to be in places with regular wind.

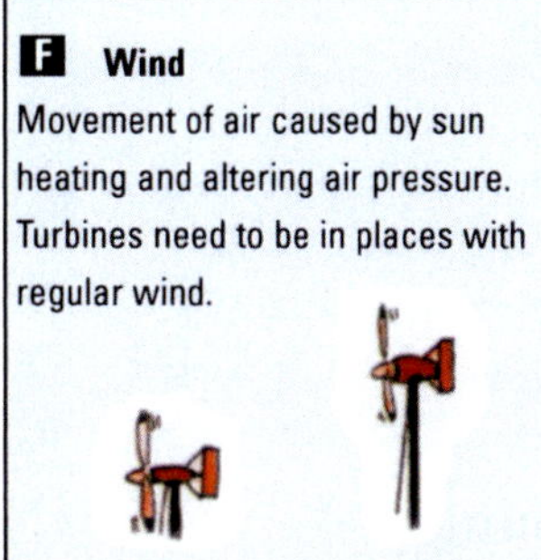

G Wave

Energy from vertical wave movement. Very costly to design and build.

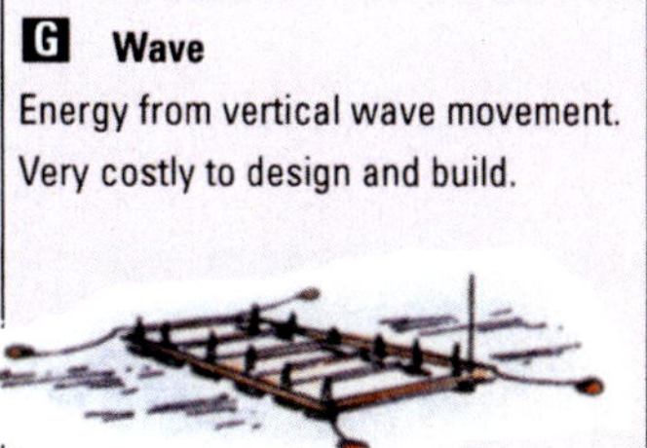

H Wood

Burned for warmth and cooking.

I Organic waste conversion

Power from gases produced in rotting organic wastes.

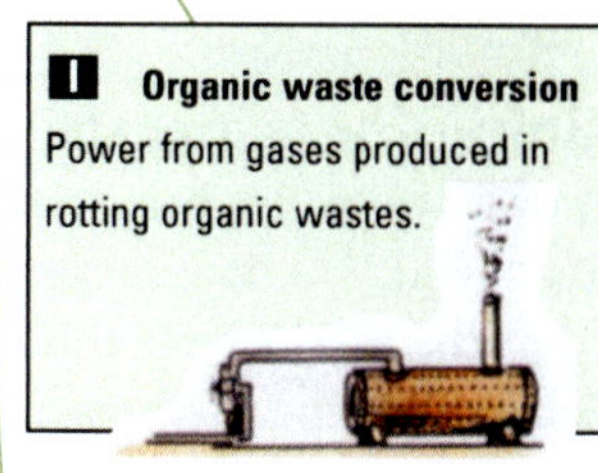

Fossil fuels

Formed from organic matter buried and compressed over millions of years. Release pollution into atmosphere.

J Coal

K Oil

L Natural gas

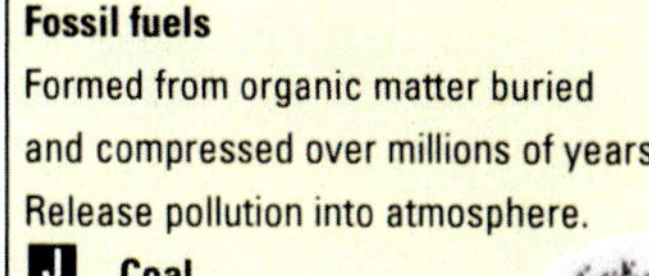

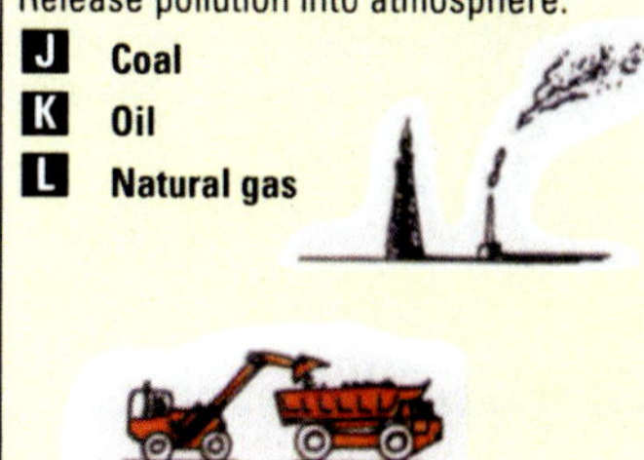

Activities: Analysing energy use

1 Measuring world energy use

a Plot the following data on a line graph. (Put units on vertical axis.)

Year	Units (billion btu)	Year	Units (billion btu)
1970	207	2000	410 (est)
1975	243	2005	449 (est)
1980	285	2010	500
1985	311	2015	544
1990	347	2020	608
1995	380		

b Explain the trend in energy use to the present.

c Why is this happening?

d What is likely to happen in the next 20 years?

2 Sources of energy

a Look at the diagrams and photos in this unit, of the possible sources of energy. Draw up a chart with four columns, using these headings:

A Energy source

B Is it sustainable? (Will it get used up or is it renewable?)

C Its advantages. (Is it efficient, easy to build, etc.?)

D Its problems. (Is it costly, will it pollute, is it dangerous, etc.?)

Energy source	Sustainable?	Advantages?	Problems?
A B etc			

For each energy source mentioned fill in the chart. You may need to discuss some of the items with other members of your group.

b **A**, **B** and **C** are not directly from the sun. Explain their origin and why each is not likely to become more important.

3 Sustainable energy

a What sorts of climate would suit the use of solar energy?

b New Zealand is well suited to the use of wind energy. Why?

c New Zealand spends a great deal of money in prospecting for oil or natural gas, but little on wind power research. Explain.

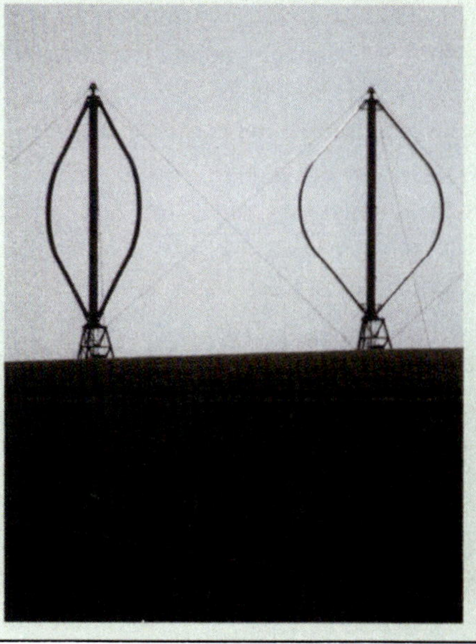

17 Non-renewable resources

As the world's population increases, and more people want to share in the riches of the wealthy countries, more pressure is being put on the world's resources. Some of these resources are **non-renewable**. Once used they are gone and cannot be replaced. These include land for farming and especially minerals from the earth.

Mining for minerals

All over the world there is pressure to find and mine minerals such as iron, nickel, copper, gold, uranium and many others. Wherever this takes place there is usually **conflict**. Are the advantages of mining worth the problems it causes?

What are the reasons for mining?

1. The mineral can be used as a **resource** to benefit people – from building the Sydney Harbour Bridge, to making jewellery, to providing nuclear fuel.
2. The mining provides **profit** for the mining company and the country.
3. **Jobs** are provided for the local people in the mine or processing plant.

What are the problems that mining causes?

1. Because minerals are **non-renewable**, they will eventually be used up and will not be available for future generations of people.
2. Mining damages the **environment**. Almost always mining and processing leaves damage in the form of ugly or useless mine sites and waste dumps and the possibility of pollution in waterways or the air.
3. It often causes **conflict** with local people – particularly indigenous (native) people whose way of life and values are affected by the mining operation.

All around the world these sorts of problems follow large mining operations. The case study of uranium mining is just one example.

In New Zealand, gold mining has had its share of conflict, especially at the Martha Mine at Waihi. The profit it makes and the jobs it provides have to be balanced against environmental concerns about the huge hole, the large tailings dump and possibly harmful wastes entering the waterways.

Martha gold mine, Waihi.

Case Study:
Uranium mining at Kakadu

East of Darwin in the Northern Territory of Australia is the Kakadu National Park. It is a beautiful wild area with spectacular landforms and varied wild life. The environment is so special that it has been declared a World Heritage Park. For at least the last 20,000 years it has also been home to the ancient Kakadu people. In the area are many ancient rock drawings and other sites sacred to the Aboriginal people. It also has some of the richest uranium deposits in the world. This has lead to conflict:

The Uranium mine and its value:

In 1980 an agreement was reached with the local Aboriginal people, and the land was leased to a mining company which began mining the uranium at the Ranger Mine, within the Park. Since then 3-4000 tonnes of uranium oxide concentrate has been produced every year. This makes Australia the world's third largest producer of uranium, supplying 8 per cent of the western world demand. It is almost all exported overseas to fuel nuclear power plants – mostly in Japan. This industry earns over $70 million each year for the company and for Australia. The company is at present attempting to set up another even bigger uranium mine at Jabiluka about 20 km away, but this plan has met with so much opposition from the local Aboriginal people and from other Australians that it has been put on hold until 2005. The mining company spends a great deal of time and money in planting and restoring the land that has been mined.

Opposition to the mining:

A The Kakadu People: They were put under great pressure to allow mining and have been given large sums of money – over $145 million so far. (They receive 4.5 per cent of the profits plus $200,000 per year.) While this money has helped the tribe in many ways it has also helped to break down their old values and disrupted their way of life. To the ancient Aborigines the land was their life, and they felt that their role was to look after it and all the spirits that are part of it. The Kakadu people know that with the loss of these beliefs their aboriginal culture has tended to break down and the people have lost their sense of purpose and belonging. The Mirrar people at the site of the planned new mine at Jabiluka, have also seen what has happened, and most are very strongly opposed to the new mine.

B Environmental concerns: Many people around Australia support the Aboriginal people and are also concerned about the dangers of mining uranium. A particular worry is that water polluted with radioactive waste may escape from the site during the wet season.

Activities: Local issues

New Zealand is lucky that we do not have severe population pressure and our resources are less under threat than in many parts of the world. But we are very wasteful of our resources, and there are ways that we can learn to conserve in our homes. Here are some:

A Conserving our energy resources

1 Home energy

Match up the 'home energy uses' items with the 'ways of saving energy', from the lists:

Home energy uses	Ways of saving energy
1 Cooking	a Fill kettle from cold tap
2 Running hot water into basin	b Use shower rather than bath
3 Water heating	c Microwave (uses 30 per cent less energy)
4 Washing yourself	d Use toaster rather than grill
5 Clothes or dishwasher	e Fluorescent bulbs use 20 per cent less power and last 10 times longer
6 Lights	f Use only when you have a full load
7 Making toast	g Keep plug in the basin
8 Refrigerator	h Keep thermostat below 60 degrees C
9 General power use	i Keep door shut
10 Boiling kettle	j Switch off if not being used

2 Saving energy at home

At home discuss the ideas on this page about saving energy. Try and arrange with your family to have a month's trial using as many of these energy-saving methods as you can. Check your power bill before and after the trial. How much money have you saved?

B Conserving our water resources

1 Home water use

When Auckland had a severe water shortage in 1994 it was found that an average family of four used 720 litres of water per day, as follows:

	Av. family use (litres)
Flush toilet	220
Showers/baths/etc.	160
Watering garden, car, paths, etc.	140
Clothes washing	120
Kitchen/food preparation	60
Leaks/taps dripping	20
Total	720

This is over 70 large buckets. (One large bucket is about 10 litres.)

a For each of the following work out how much water you use to: flush the toilet, run bath or shower, wash yourself, wash dishes, wash clothes, prepare vegetables, drink, wash teeth, any other water-using activity.

b For each work out how often it is done each day in your home.

c Total your home water use. How does it compare with the list above?

2 Conserving water

a From your water meter in the street read your water use for one week.

b Arrange with your family to do a water conserving trial for a week. (Use the ideas listed.)

c At the end of the week read the meter again. How much did you save? (Careful use could save 200 litres per day.)

Home energy saving ideas

Cooking

1 Use a shiny reflector (foil) under hot plates on an electric stove.
2 Keep the lid on pots while cooking.
3 Use several pots that fit together on one element.
4 Thaw out frozen foods before starting to cook them.
5 A fan oven is more efficient.

Space heating

1 Glass is a poor insulator – use curtains to keep heat in.
2 Close off unused rooms in winter.
3 Insulate ceiling (can repay costs in as little as 1–2 years).
4 Put a wrap around the hot water cylinder. (An $80 wrap should save about $65 per year in electricity costs.)

Hot water

1 Repair dripping hot water taps.
2 Put wrapping around hot water pipes.
3 Install low-flow shower head.
4 Gas or solar hot water heater makes big savings.

for more check
www.eeca.govt.nz/content/tips

Activity: Conflict at Kakadu

1 Locating the issue

On a small map of Australia mark in Darwin, Kakadu National Park and the Ranger Mine site.

2 Clarifying the issues

Below the map write sentences explaining:

a The viewpoint of the mining company.
b The benefits of mining to the aboriginal people.
c The disadvantages to the aboriginals.
d The viewpoint of the environmentalists.
e Your viewpoint about the mining.

3 Roleplay

Within your group or class, arrange to role play a meeting:

A between the mining company and aboriginals to sort out the conflict.
B between young aboriginals and the elders.

You will need to appoint a chairperson to keep the discussions peaceful!

Ways to save water in the home

1 Fix all washers.
2 Don't leave tap running when cleaning teeth.
3 Sweep paths instead of hosing.
4 Only wash with a full load.
5 Don't overwater garden plants.
6 Don't run water to wash the spuds.
7 Don't leave hose running when washing car.
8 Shower with low volume head.
9 Install a dual flush loo.

The waste mountain

18 Polluted water

Fresh water is vital to humans. It was shown on page 24 that it is rapidly becoming scarce, but even worse, what water we have is often so polluted that we can not use it. It has always seemed so easy to get rid of waste by dumping it into a river or the sea. We thought that rivers just carry it away to the sea, which is so large (covering 70 per cent of the Earth's surface) that it would cope forever. We are now finding that this is no longer the case. Most waterways are now polluted by our waste. In most parts of the world it is not safe to drink from a stream without first treating the water. Even in New Zealand, with its clean, green image, we can no longer drink safely from streams in the countryside as our grandparents' parents could. What have we done? Here are some reasons why our water is no longer pure – and what we have to do about it:

Causes of water pollution

1. Pesticide sprays
2. Fats, oils detergents
3. Factory poisons that spill into drains
4. Topdressing with fertilisers
5. Toxic chemicals from factory wastes
6. Mammals with parasites
7. Chemical wastes from factories
8. Animal wastes
9. Sand and silt washed from developments
10. Organic (food) wastes
11. Plastics and hard solids washed into drains
12. Forest removal
13. Harmful micro-organisms from food wastes and sewage
14. Run-off from roads – rubber, poisonous metals from exhausts
15. Household sewage
16. Run-off from roofs and gardens
17. Metals and other solids from factory wastes

Pollution of rivers and lakes in the countryside

Our ancestors' clearing of the forests from the countryside and replacing them with farming has led to pollution of rivers and lakes in ways that they never dreamed of:

1. There is more erosion, rivers **flood** more quickly and carry more **silt**.
2. The rivers and lakes have more **nutrients** pouring into them, which makes water weeds grow more quickly.
3. Some rivers contain **poisonous chemicals** from agricultural sprays.
4. Tiny **harmful micro-organisms** such as Giardia that live in the gut of mammals are present in many streams. They cause unpleasant stomach upsets in humans.

Because most of us now live in cities, the waterways around them become a very concentrated source of pollution. Water from the larger towns and cities is such a pollution problem that the larger city councils have to spend a great deal of time and money dealing with it. There are two types of problem:

Wastewater

This is water that leaves our homes containing large amounts of wastes – from the kitchen, laundry, bathroom and toilet. It also takes harmful wastes from factories. This wastewater is very polluted and in most cities is collected in a city sewerage system and taken to a wastewater treatment plant. Here the sludge is separated from the water, which is purified to various levels before being released into the sea or waterways.

Did you know that there are still several large cities in New Zealand that do not have waste treatment plants, and release untreated waste into the sea? These cities include Dunedin, Wanganui, Napier and Hastings.

Stormwater

This is the water that falls on the city as rain. Because so much of the surface of cities is covered in roads, roofs and paths that do not absorb water, it must be collected into stormwater drains to prevent it flooding the town. These drains take it to nearby streams or the sea. Unfortunately the stormwater tends to collect up many other things so that when it reaches the outlet it may be quite polluted. It may contain rubber and poisonous metals from the roads, harmful chemicals from factories and silt from developments. Many towns spend large sums of money trying to keep stormwater as pure as possible and separate from the wastewater drains.

The sea – oil spills

The sea is used as a dumping ground for many things, but some of the worst pollution problems have been accidents – especially oil spillages from the huge supertankers that carry oil fuel around the world. These enormous tankers may be the size of several football fields, and may carry over 130,000 tonnes of crude oil. Because of their size they are difficult to manoeuvre and there is always the possibility of an accident.

The Exxon Valdez disaster

One of the largest and best known of these accidents took place in March 1989.

- The tanker Exxon Valdez was moving out of Puegot Sound in Alaska with a load of crude oil when it ran aground.
- 41,000 tonnes of crude oil leaked into the sea, and winds and tide carried much of it to the coast. Nearly 1900 kilometres of coast were coated with a stinking sticky mess of oil.
- Over the next year the Exxon Oil Company spent more than $1.5 billion trying to clean up the mess. Over 11,000 people and 1400 boats were involved.

Cleaning up oil spills

This is a terrible job because:

1 Heavy oil forms a thick sticky mess that is hard to pump or scoop up.
2 Hot and cold water sprays are used to force oil off the rocks, but each tide is likely to bring a new layer of oil to the beach.
3 Floating booms are used to try and enclose the oil which is skimmed up in calm weather.
4 The feathers of seabirds stick together and lose their waterproofing. Most die. It takes several hours and much water and detergent to clean each bird, which still may die.
5 Workers find the job very dirty, very tiring and very frustrating.

Activities: Finding out about water pollution

1 Causes of water pollution

Rule three columns in your book labelled:
A Rivers and lakes; **B** Wastewater;
C Stormwater.
Put the 17 different forms of water pollution in the chart on page 32 into the columns where they are most likely to be found.

2 Publicising water pollution

Draw a poster showing people how they pollute our waterways. (Use the information and diagrams in this unit.)

3 Where does the water go?

a Check out and note where the stormwater from the roof of your house goes.
b Find out and note where the wastewater from your kitchen and bathroom goes. Is it different from the stormwater outlet?
c If you are in a city, the outlets should be different. Explain why?
d What happens to the wastewater in the city nearest to you?

4 Visualising an oil spill

Supertankers the size of the Exxon Valdez bring crude oil to the refinery at Marsden Point near Whangarei about three times a week. Imagine a tanker ran aground on rocks near the Whangarei Heads and 30,000 tonnes of oil spilled out.

a Draw a map of New Zealand. Mark in the refinery at Whangarei.
b If on the same scale as the Exxon Valdez spillage, over 1000 kilometres of coast would be polluted. This is more than from North Cape to Coromandel. Find these capes and shade in pink all the polluted coast in between.
c Very heavy pollution might reach over 80 kilometres of coast – say from the Whangarei Heads to Kawau Island. Find these places and mark this area in red.
d 10,000 people are needed to work for a year cleaning up. What proportion of Auckland's total work force of 500,000 is this?
e The cleanup cost is $1 billion. What proportion of New Zealand's whole education budget of $4 billion is this?
f Imagine you have just spent a week working on cleaning up the coast. Write a letter explaining what the job was like.

5 Research on water treatment plants

Do some research to find out about the water treatment plant in the town nearest to where you live:
a Where is it?
b How does it work?
You may be able to get information from your local city council. If not, look at how Christchurch city deals with its waste: website: www. ccc. govt. nz

19 Polluted air

In the northern hemisphere, where most people live, air pollution is a serious problem, especially in the large cities. Here's what it can be like:

1 In the Chinese industrial town of Lanzhou, on a still day your eyes sting and the back of your throat prickles from the air, polluted by the haze made up of fumes from the coal-burning power plants and other heavy industrial factories.
2 In Los Angeles, 15 million people breathe the worst air in the United States. 1,600 people die due to smog each year, and many thousands more suffer from respiratory illnesses.
3 Often Delhi, in India, is covered in a thick blanket of smog. Even the opposite side of the street is hazy, and the policemen wear gas masks on the streets.
4 In some Japanese towns there are oxygen booths where people can pay money to breathe some oxygen enriched air.

Smog

Most of these examples are of smog. This forms over cities and industrial areas for several reasons:

1 Some countries burn coal and oil in large amounts for power and industry, and this releases large amounts of sulphur dioxide. (Over 100 million tonnes/year!)
2 Burning petrol or diesel fuel in vehicles releases carbon compounds, nitrates, lead, dust and other chemicals. These make up high concentrations of small particles (called aerosols) which are the cause of a brown haze over cities.
3 When the air is still for a few days the sun causes these chemicals to form smog – made up of ozone and other harmful chemicals which form a thick haze.

Acid rain

This acid rain problem has been largely brought under control in western Europe, but is still a big problem in Russia and eastern Europe. It is so serious that:

1 70,000 square kilometres of European forests have been destroyed by acid rain (one third of the area of New Zealand).
2 Many famous stone buildings are crumbling away from the acid in the rain.
3 The water in many thousands of lakes in Northern Europe and America is so acidic that no fish can live in them.

The diagram shows how acid rain is caused:

Thick smog makes life difficult in Mexico City

MEXICO CITY A thick smog covered Mexico City on Thursday, causing widespread illness and financial damage.

The longest-running pollution alert period since 1996 was extended after officials banned two-fifths of all Mexico City's cars from the streets for the fourth straight day. Hospitals have reported sharp increases in the number of breathing difficulty cases, as well as high numbers of migraine and eye irritation cases.

Ozone pollution at 3 p.m. registered 194 on Mexico City's air pollution index, almost twice the level considered safe...

Schools banned outside activities and residents were advised to stay indoors. Despite the warnings, health problems have continued to increase over the last 10 days.

Almost a third of Mexico City's population is at risk of health problems caused by the air pollution crisis. The Mexican Health Ministry has documented 7.2 million visits to doctors and hospitals this year for respiratory ailments – 1.2 million more than in the same period in 1997.

CNN 29 May 1998

Acid rain – causes

A Sulphur and nitrogen compounds from burning coal and oil in industry.
B Nitrogen oxides mainly from vehicle exhausts.
C Mixed with water vapour it forms a 'soup' of weak sulphuric and nitric acids.
D Dry particles and acidic gas also carried in air.
E Tall chimneys cause pollution to be carried great distances – even into other countries.

Acid rain – effects

F The acidity of freshwater lakes and streams increases, eventually killing the fish.
G The surface of stone buildings slowly crumbles and metal corrodes away.
H In soils it reacts with other minerals concentrating some and leaching out others.
I Trees die, especially in naturally acid soils.
J Soil water is acidified, causing harmful metal levels to build up.
K Health problems affecting breathing increase.

What about New Zealand?

Many of the northern hemisphere countries are spending a lot of money trying to reduce air pollution. In New Zealand we are used to clean air and blue skies. But even here there are pollution problems – mainly in Christchurch and Auckland.

Vehicle pollution in Auckland

Auckland is both warmer and windier than Christchurch, and because of this it has fewer air pollution problems. But on calm winter days a brown haze appears – a sure sign of pollution. It is caused by:

Motor vehicles	77%
Domestic fires	16%
Industry, etc.	7%

Motor vehicle pollutants that cause most of the problems are:

Carbon dioxide – a gas produced in huge amounts. It helps heat the air and leads to global warming (see page 42).

Carbon monoxide – a poisonous gas that causes headaches, dizziness and nausea.

Nitrogen oxide and hydrocarbons – gases that irritate the lungs and affect asthmatics. They also smell and help form brown haze and smog.

Poisonous particles such as lead – form a layer over roadside plants, etc.

Christchurch smog on a cold winter morning.

Smog in Christchurch

The worst air pollution problems take place in Christchurch each winter. Smog levels rise above danger levels on about 30 days each winter because:

a In winter large amounts of **suspended particles** are released into the air – almost 90 per cent from solid fuel burners used to heat houses.

b Christchurch is low lying and damp in winter and on still, cold frosty nights a **temperature inversion** can take place. Suspended particles in the air are trapped near the ground under layers of warmer air.

Smog levels dangerous to health are calculated as more than 50 ppb (billion particles per cubic gram).

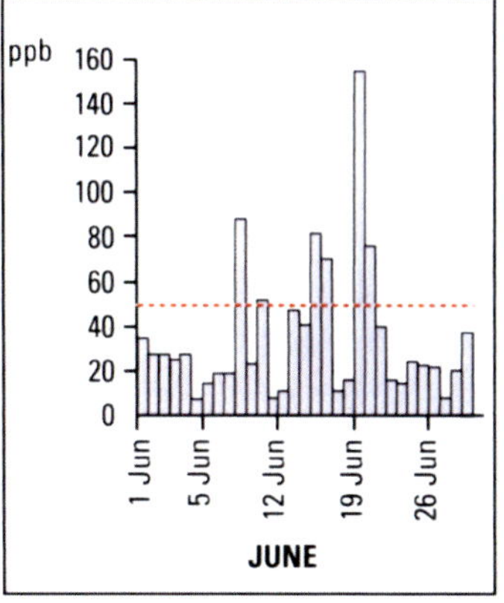

Air pollution readings in Christchurch for June 2000.

Activities: Studying air pollution

1 Interpreting Mexico City news item

Answer these questions:

a What is the air pollution index level considered in Mexico to be safe?

b About how much above this did the level reach?

c List three health effects of the pollution.

d What proportion of the city population is at risk?

e What was done to reduce the pollution on smoggy days?

f Did it seem to be working?

g Mexico City is in an inland basin and has little wind. How would this affect the pollution?

2 Smog

a List three ways that you may be affected by breathing heavy smog.

b Sulphur dioxide, nitrates and lead particles may all be in smog. For each note where it comes from.

c List two weather conditions that help to produce smog over cities.

3 Acid rain

Copy the acid rain diagram and put all the labels in alongside their correct numbers.

4 Data on Christchurch smog

Average concentrations over a smoggy day (ppb = parts per billion)			
9am	50ppb	12am	300
12pm	10	3am	220
3pm	10	6am	110
6pm	175	9am	100
9pm	370		

a Plot this data on a line graph. (Put time on the horizontal axis.)

b What proportion of the samples were above the danger limit?

c When did the smog build up?

d Explain why.

e From the graph for pollution levels in June note how many days had pollution above safe levels.

f A large anticyclone passed over the South Island during 15–21 June. What effect did this have? Explain why.

5 Motor vehicle pollution

a Draw a pie diagram showing the sources of air pollution in Auckland.

b Cars that are idling produce four times as much pollution. Why is this a particular worry in Auckland?

c List three health affects of vehicle pollution.

d List two other effects.

6 Promoting less car pollution

Imagine you are promoting a campaign to either use public transport instead of cars or make cars less polluting.

a Design a suitable logo.

b Write some catchy campaign slogans.

20 Rubbish mountains

Imagine Eden Park filled with a pile of rubbish five stories high. That represents the rubbish produced by Auckland in one month. If we also added the waste produced through all of New Zealand in one month we would more than fill Jade Stadium in Christchurch and Carisbrook in Dunedin!

We are a society of waste makers. Although poor countries often seem untidy, the proportion of waste produced by each person is only a small fraction of what a society like ours produces. The wealthier the country the more solid waste we produce. Dealing with it is a special headache.

Where does the rubbish come from?

New Zealand produces well over 3 million tonnes of waste each year. It comes from two sources:

Household wastes: The average household produces over 2 kg of waste every day. This is 750 to 900 kg per year! A look in a typical rubbish bin reveals what most of it is made up of.

Commercial waste: Over 60 per cent of all waste comes from businesses. Commercial firms produce particularly large amounts of paper and cardboard. Metal wastes come from industries, and building sites produce a great deal of construction and demolition waste.

The waste from a typical household is made up as follows:

50%	kitchen and garden waste (organic)
25%	paper (newspapers, magazines, wrapping, etc.)
10%	plastics
5%	glass (bottles)
5%	metal (cans)
5%	other

How to deal with it?

Of course most waste comes from the cities, and it is the councils in the big cities that have the problem of dealing with it. Auckland has the biggest problem, and has to deal with nearly one million tonnes of waste each year. Until the 1980s most cities simply carted it to huge rubbish tips but this is becoming a more difficult option. Here are the choices:

Waste from our consumer society

Reduce –
Reuse –

Generally the amount of waste being produced per person is still increasing, and new landfills are becoming harder to find. So some councils are looking at ways of getting people to reduce the amount of waste. (See page 38.)

Recycle –
Recover –

Many cities have operated recycling schemes for some years. Some ask people to put out paper, glass, plastics and tins separately. Sometimes recyclable waste is separated at a transfer station. But recycling is still quite costly and needs mass cooperation. (See page 39.)

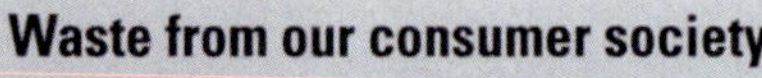

Disposal

Most waste still finishes up in landfills. In Auckland about 30 per cent of waste is recycled and three-quarters of the rest could be recycled. This would mean landfills would last four times as long (perhaps 80 years instead of 20!) Why not burn it? Incineration is too costly and polluting to consider. Only medical and quarantine wastes are burned in incinerators.

Love Canal – a landfill gone wrong

Love Canal is a landfill site in Niagara City, USA, that was used in the 1940s and 50s by a chemicals company to get rid of wastes. In 1953 the site was covered and a school built on it. Twenty-five years later it was found to be leaking so much toxic waste that the school had to close and 950 families had to be evacuated from their homes. Poisons were seeping into the water supply and harmful wastes and gases were being released.

Since then several hundreds of millions of dollars have been spent in cleaning up, and over $14 billion filed in lawsuits against the chemicals company. Cleaning up dumps like this is a nightmare, yet there are perhaps 50,000 hazardous waste sites in the USA alone.

Fortunately we have learned from stories like this, but disposal of wastes remains a problem.

Landfills

All towns have a rubbish tip. Usually it looks untidy and, especially in small towns where tip management is too costly, it may be a health problem. In large cities locating and managing large landfills are big problems.

Choosing a landfill site

This is a big problem because nobody wants one near them. How would you like to live near a tip, with its smell, noise, dust, seagulls, rats and possible health problems? It would also mean that nearby property values would drop a good deal.

Auckland, with its huge amount of waste, has five landfills. Two will be full by 2005 and a new site is being looked for. A good site needs to be:

1 Large enough to cope for many years.
2 Close enough to keep transport costs low.
3 Be on land of low value for farming.
4 Far enough away from settlements not to cause health, smell, dust or noise problems.
5 Able to be landscaped and developed after being filled.
6 Have soil conditions that do not present seepage problems.
7 Have no Heritage and Maori sites that should be preserved.

Managing a landfill

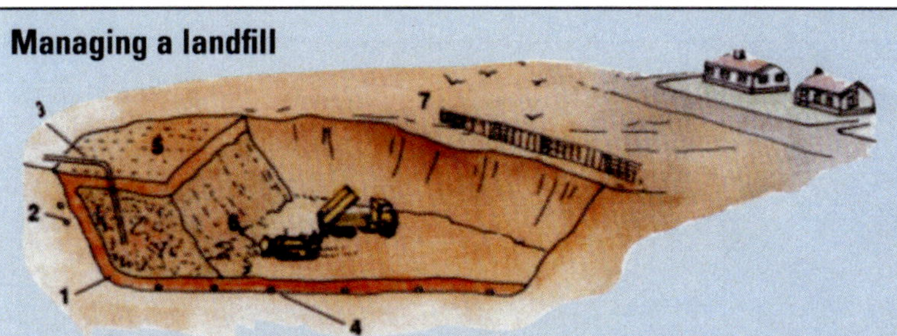

1 Sides and base need to be sealed with clay to stop liquid wastes seeping out.
2 Drains needed to control water levels.
3 Liquid wastes (leachate) drained off from base and taken to sewage plant.
4 Gases generated from the decaying waste collected and piped away for use.
5 Surface sealed with clay and soil and then landscaped.
6 Tip face managed by rapid burial of waste.
7 Seagulls and rats controlled. Dangerous chemicals monitored and kept as low as 0.02 per cent of the waste.

Activities: Sorting the rubbish

1 Domestic rubbish

a Draw a pie diagram to show the proportions of different wastes in a week's rubbish for an average household.

b How does this compare with your household's weekly rubbish? If you are really keen you could tip it out onto newspaper and divide it up (use gloves). You may be able to get rough percentages just by carefully watching what goes in. How do your figures compare with the average figures? Explain why.

2 Analysing the sources of waste

Commercial waste varies, but in Auckland it is made up of the following proportions:

building/demolition	28%
metals	23%
paper	20%
organic	15%
plastics	9%
glass	1%
other	4%

a Draw a pie diagram of this data.

b Compare the pie diagrams for domestic (question 1) and commercial waste. List the two main differences.

c Explain these two differences.

d If everybody recycled all their organic and paper waste, how would this affect the amount of rubbish?

3 Dealing with the waste

a Explain why landfills are becoming a difficult option for waste disposal.

b Give two reasons why burning all the waste is not a good option.

c Why did Love Canal become a disaster? Who was to blame? Why?

4 Understanding the terms

Write down the meaning of the following terms (refer to pages 36 and 37):

leachate, toxic, incinerate, quarantine.

5 Class discussion – siting a landfill

Imagine that there are plans to site a landfill in a quarry in your suburb. Set up a class discussion to consider the issue. Use ideas from this unit. Set up three groups:

Group 1 – is a council committee that listens to the evidence, asks questions and makes the decision as to whether to use this site for a landfill. One person will be chairperson of the discussion.

Group 2 – are members of a waste management company who are looking for a site for a new landfill and feel that the quarry is the best site. They must present their arguments.

Group 3 – represent the 50 house owners who live within half a kilometre of the quarry and who do not want a landfill there. They must present their arguments.

21 Dealing with the rubbish

Reducing and reusing waste

Many New Zealand towns have been looking to recycling schemes to try and deal with waste. Often this is difficult, particularly in smaller towns where amounts of waste are small. Recently more effort is being made to persuade people to actually produce less waste. Two examples are:

a **Education** – several cities have been running publicity campaigns to make people more aware of the waste problem (see the box).

b **Cost** – Auckland City has replaced all large wheelie bins with small bins.

In some poor countries, such as Cambodia, rubbish tips are badly managed, and many people survive by scavenging in the tips for odd items that they can sell or eat! What are the hazards for the rubbish tip people?

Ways of reducing waste

Around the shops:

1 Take your own shopping bags and say no to plastic bags.
2 Don't buy items with wrapping that is not needed.
3 Avoid throw-away items – paper and foam cups, plates, disposable razors, etc.
4 Reduce packaging by buying in bulk.
5 Buy products made from recycled materials where possible.
6 Choose soaps and detergents that are biodegradable (will decay away).
7 Buy long-lasting metal kitchen utensils.
8 Shop locally and use public transport where possible to save on fuel and prevent pollution.

Around the home:

1 Reuse glass or plastic containers if possible.
2 Use kitchen scraps as compost.
3 Take old books, magazines and toys to hospitals, schools, etc. where they can be reused.
4 Save clothes for clothing bins or opportunity shops.
5 Sort and take part in recycling schemes in your area.

Activities: Finding out about reducing waste

1 Ways of reducing waste

In your groups discuss the different suggestions for reducing waste. In each case decide:

- Why the idea is suggested.
- How practical the idea is.
- If your household is doing this or not.

2 Recycling poster

Imagine your group is a committee set up by the local council to encourage recycling. Design a poster, a logo and some slogans that could be used in a publicity campaign.

3 Recycling schemes

a Draw a pie diagram (see page 39) of the proportions of different kinds of waste recycled in Auckland.
b Write sentences explaining how this differs from the amounts of waste figures from the previous unit.
c What type of waste recycling could be improved the most?
d List three reasons why plastic waste is an especially difficult problem.
e List four reasons why recycling is difficult.

4 Recycling in your town

For your local town, find out:

A What the local council's present recycling arrangements are.
B What plans the council has for reducing waste and recycling in the future. (Your local council may have a website.)

5 Composting

Much of your school rubbish is organic matter. Plan a composting system for your school. Find answers to:

A how to organise collecting and sorting the organic matter,
B placing, designing and making a bin,
C looking after it,
D what to do with the compost.

Perhaps you could persuade the school to set up a composting system. Or you could set one up for your own home kitchen wastes.

6 Attitudes to reducing waste

Look at what all these people are saying about the suggestions for reducing waste. For each one, decide if it is a sensible comment, then write down your reply to it. (You could discuss these in your groups.)

A Rubbish is not my problem – I pay rates for the council to look after it.

B Why bother, my efforts would make no difference.

C It is too embarrassing saying 'no' to extra wrapping at the shops.

D Sorting rubbish for recycling takes too much time.

Recycling and recovering waste

Recycling seems sensible and not too hard. Yet still only 30 per cent of our wastes are recycled. Why is it so difficult? There are two main problems:

1 **Cost**

a **Sorting** – Separating the different kinds of waste is very time consuming and expensive. It helps if some is done by the household.

b **Transport** -The recycling plant may have to collect material from a wide area to get enough to make it worthwhile.

c **Processing** – Processes such as cleaning, crushing, separating and refining may be more costly than making new items.

d **Finding markets** – The cost or quality of the recycled material may mean that the industry may not be able to find buyers for it.

2 **Lack of interest**

Many people can not be bothered getting involved – or often they have never thought about it.

What is recycled?

The 30 per cent of waste recycled in the Auckland region is made up as follows:

Metals	37%
Paper/cardboard	45%
Compost	9%
Glass	4%
Plastics	3%
Aluminium cans	1%
Clothing/rags	1%

Many cities try to reduce the cost of recycling by asking households to do the first sorting. In Auckland paper should be tied up separately, and plastic, cans and bottles put in a separate container. Not difficult, but only about 60 per cent of households take part.

Plastics

Plastics make up about 9 per cent of our wastes and have always been a problem because:

- Most plastic is not biodegradable so it becomes long-lasting litter. However, some newer plastics, such as 6-pack tops, are photodegradable – will break up in sunlight.
- Nylon fishing nets and plastic wastes in the sea harm marine life.
- It gives off poisonous fumes when burnt.
- There are many types of plastic and these need to be separated before recycling.

Some cities now separate plastic at a sorting depot. After sorting it is shredded into thumbnail-sized pieces which are sterilised, dried and bagged. Much of this is sold overseas, but some will be recycled in New Zealand – perhaps as clothes pegs, or even milk bottles again.

Metals

The most successful recycling items are metals, because they are too valuable to throw away. Most towns have a scrap metal depot or scrap dealer. Most metal wastes come from businesses.

Iron – Recycling of scrap iron is most important – by Pacific Steel and NZ Steel.

Other metals – Metals such as brass, copper and lead are melted down and remade into new products.

Aluminium – Aluminium is widely used for cans, foil, carton tops, etc. Although it must be cleaned and sorted, when recycled it only uses 5 per cent of the energy needed for refining. Many schools have been involved in Comalco's aluminium can recycling scheme. Comalco donates money from this scheme to the programme to save the Kakapo.

Paper

All cardboard and paper can be recycled unless it is wax or plastic coated. This is the second most successful form of recycling in New Zealand and several cities have collecting and recycling schemes. Most is made into cardboard and cheaper products such as egg cartons. However the market for recycled paper is still fairly small.

Organic matter – (kitchen and garden wastes)

Generally this is not only the biggest part of household waste, but it is also the part that could be reduced the most.

It includes all kinds of organic matter – waste food, peelings, even paper. It can all be composted to enrich the soil for growing crops.

Some cities do have plants which use organic matter from rubbish collections to make compost on a large scale.

Glass

Glass is ideal for recycling and glass recycling schemes have tended to be quite successful. Once it is sorted and crushed it can be added directly to the molten glass in the glass-making process. However, it still costs 2c to recycle a bottle and 1.9c to put into a landfill. So greater volumes of glass are needed to make recycling cost effective.

Other wastes

1 **Clothing and rags** – Most towns have collecting depots for old clothes.

2 **Car batteries** – These can be recycled to obtain the lead.

3 **Used oil** – This can be cleaned and reused.

4 **Tyres** – Used tyres are a problem. In New Zealand, only a little rubber is ground up and reused.

22 Dangerous wastes

Amongst all the wastes humans produce are some that cause very dangerous problems. These are mostly chemicals that are toxic (harmful) to humans. They cause several kinds of problems:

1 **Toxic waste dumps:** Through ignorance or carelessness toxic wastes have been disposed of in ways that may later cause problems. There are over 50,000 toxic waste dumps in the USA alone.

2 **Radioactive waste disposal:** Because radioactive wastes may not decay for hundreds of years it is difficult to know what to do with them. Even moving them from one place to another is dangerous. Some nuclear wastes have been sealed in concrete and dumped in the sea. Others have seen stored in deep mines.

3 **Accidents:** These take place for many reasons – faulty equipment in the factory, carelessness (e.g. Chernobyl), lack of knowledge, etc.

4 **Pesticide problems:** These are poisonous chemicals used for destroying farm pests. Early pesticides such as DDT were not biodegradable. As a result they built up in the soil and in plants. When eaten, they could become so concentrated in animals in the food chain that they caused harm.

5 **The wastes of wars:** The most horrifying of all wastes are those released on purpose by humans in wars. The atomic bomb dropped on Hiroshima in 1945 killed over 80,000 people, and in the last 20 years land mines have killed or maimed many hundreds of thousands of innocent people.

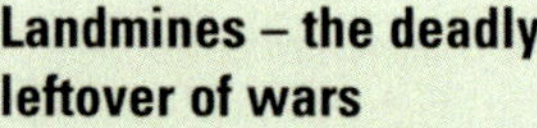

Landmines – the deadly leftover of wars

- Think about this: *somewhere in the world an innocent person is killed or maimed by a landmine every 20 minutes or so* – that is about 26,000 people every year.
- This is because all around the world, wherever there has been a war, landmines that have been buried in the ground are left behind, their position unmarked. There are thought to be about 100 million unexploded landmines still buried in about 60 countries.
- The cost and danger involved in finding and removing all of these is huge – far beyond the ability of many of the countries involved. It makes one wonder about the stupidity of the human race!

Some examples of dangerous waste problems

A Chernobyl

This is a name that worries all environmentalists. It is the site of the single most serious chemicals disaster caused by humans. These are the details:

- Chernobyl is the site of a huge Russian nuclear power plant.
- In April 1986 an accident took place while the safety systems were being tested.
- A reactor melted down and radioactive matter was released into the air.
- 31 people were killed and over 200 suffered acute radiation sickness.
- 135,000 people were evacuated from the surrounding area.
- Radiation entered the soil, water, forests and food.
- Wind spread the radiation over much of Europe – as far away as Greece, Norway and even England – over 2000 km away.
- Many people have since died of cancers caused by the Chernobyl radiation.
- Cleaning up the contaminated areas is still going on, and has already cost the Ukraine many billions of dollars.
- In 2000 the complete power plant was closed down. The United States and other European countries have agreed to pay $1.7 billion to encase the unstable reactor in a concrete shell.

B The Minimata disease

- In the 1930s a large chemical manufacturing plant began dumping waste into the sea off the little Japanese fishing village of Minimata.
- By 1968 about 27 tonnes of poisonous mercury compounds had been dumped into the sea.
- In the mid 1950s many people began to be affected by a strange disease which seriously damaged their nervous systems.
- The disease was traced to mercury which had been absorbed by the fish that were the main part of their diet.
- This 'Minimata disease' has killed over 300 people and seriously affected over 3000 more.
- At first the company refused to admit responsibility, and it has taken over 40 years of legal battles for some victims to receive compensation.

C Bhopal – escape of poisonous gases

- In 1985 an accident in a chemicals factory in India caused poisonous gases to be released into the town of Bhopal.
- Many thousands were affected by lung or eye damage. Over 3800 people died and more than 2700 are permanently disabled.
- The company paid nearly $1 billion in compensation and in 2000 it completed building a hospital for the town.
- Suspicion remains that the tragedy was caused by sabotage by an employee.

Activities: Local issues

A School litter – reducing and disposing of it

Aren't you tired of lectures about keeping the school tidy? Disposing of the bins of litter is one problem, but getting litter into them is even harder. Suppose your class is given the responsibility for planning a practical way of dealing with school litter.

- Discuss either in your groups or as a class, how you would set up a campaign:

a to reduce the amount of school litter
b to deal with the litter.

You will need to consider such things as publicity (posters, etc.), the systems you would set up (bin position, etc.) and the ways that you would make them work.

- You may be able to start a school campaign using your ideas.

B Measuring stream pollution

Measuring stream pollution in local waterways

A very useful and practical project for your class is to monitor the water quality in local streams over a period of time.

a Pre-planning is needed. You will need to sort out what stream will be monitored, where it will be sampled, for how often and for how long, and which people and groups will do the various jobs.

b Learn what can be tested and how to do the tests.

Note: Many local councils have detailed information about how to do this and may be able to help you – and in some areas may even supply testing kits. Check www.newhouse.co.nz for more details. (If this is not a science class, it may be possible for the science teacher to demonstrate the tests in a science class.)

c Sort out how the results of each test are to be recorded. Graphs should be built up of the results.

d Each recording day note the weather and any other conditions that may affect the results.

e At the end of the time, discuss the results and decide if further action is needed to improve the stream conditions.

f You may be able to discuss your results with the local council or landowner.

The following tests should be possible with a little practice:

A pH (acidity of the water)
B Flow rate
C Temperature
D Water clarity
E Water invertebrate survey

These two tests can be conducted with suitable equipment:

F Dissolved oxygen
G Phosphates
H Nitrates

Activity: Dealing with dangerous wastes

1 Terms

For the following terms write down a definition (Refer to pages 39 and 40):
Toxic, pesticide, biodegradable, photodegradable.

2 The size of the Chernobyl disaster

Draw a map of New Zealand. Imagine the Chernobyl disaster took place in Hamilton.

a Draw a circle centred on Hamilton with the radius the distance to Auckland. Shade this area and label it 'Everyone from here evacuated'.

b Shade in the rest of the North Island in a lighter colour. Label this 'Everyone here living in a contaminated area'.

c Draw an arrow to Invercargill. Label this 'Some radiation reaches Invercargill'.

3 Researching waste disasters

For each of the three examples of harmful waste disasters, A, B and C and also the Exxon Valdez (page 33) and the Love Canal (page 36) disasters, explain:

a What was the cause and who appeared to be responsible.
b Who the victims were and what happened to them.
c What was done about it.

4 Landmines – a moral issue

Discuss this issue in your groups. What is the responsibility of these groups?

a The soldiers who plant the mines.
b The directors of the armaments factories, mainly in the wealthy countries, who continue to make and sell these and other weapons to other countries.
c The governments of the countries with the armaments factories.
d The governments or groups who buy the armaments.

The future of the Earth

23 Global warming

Look at the two newspaper reports on global warming. After years of debate, most scientists now agree that the Earth is getting warmer. Although the ice ages in the past show that the temperature of the Earth has changed many times, there is evidence that the recent change in climate is at least partly caused by the activities of humans. Here are some conclusions that most scientists now agree on:

1 Average global temperatures have risen by about 0.6°C in the past 100 years.
2 Average sea levels have risen between 10–25 cm in the past 100 years.
3 Extreme weather patterns are more common.
4 *The rate of temperature change is much faster than changes ever before measured.*
5 Increased warming seems to be linked to cities and to activities that humans are involved in such as forest clearing.

The greenhouse effect

The surface of the earth is kept warm enough for humans to live on by what is known as the 'greenhouse effect'. Recently this effect seems to be increasing. This is explained in the diagram below:

Natural greenhouse effect

Heat energy reaches the earth. Some heat is reflected and escapes. Much is trapped by the 'greenhouse gases' (mostly carbon dioxide), and it returns to the Earth to warm it.

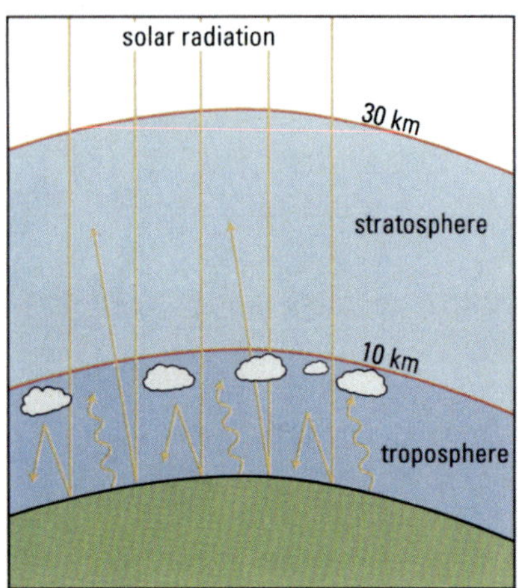

The carbon dioxide is released from the respiration and decay of living things. A little methane and nitrous oxide is also released, helping to make up the greenhouse gases. The amount of carbon dioxide is kept in balance by being reabsorbed by plants in photosynthesis and by being absorbed in the sea – some becoming part of limestone.

Effect of increased greenhouse gases

If the amount of carbon dioxide and other gases in the atmosphere builds up, clouds increase. Less heat (infrared rays) escapes and the earth heats up.

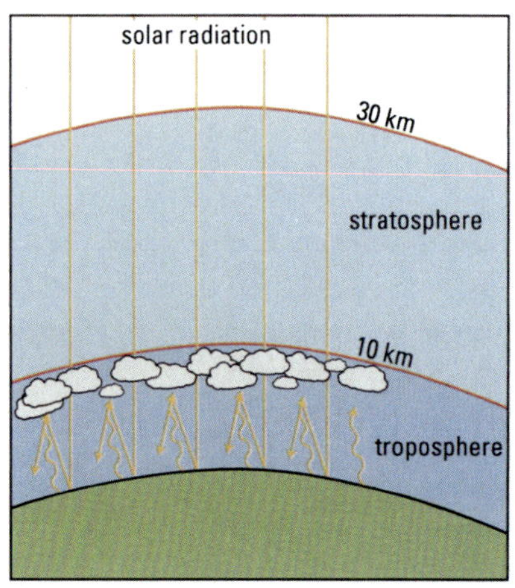

The increase in greenhouse gases – carbon dioxide, methane and nitrous oxide, caused by humans, is because of:

1 More burning of fossil fuels (oil, petrol, coal) releasing all three gases.
2 More livestock (cattle, sheep) giving off methane and carbon dioxide as wastes.
3 Removal of forests (which absorb carbon dioxide).

Global warming a reality

After decades of debate only a handful of diehards still dispute the idea that human activity is heating up the planet. All the signs seem to point that way: storms have become more intense and weather patterns more erratic; the past decade has been by far the hottest on record; and the rise in temperature has been greatest in polar regions and around cities. These facts dovetail ominously well with the theory that carbon dioxide released by burning coal, oil and gasoline for heat, electricity and transportation, is trapping excess energy from the sun. Global warming is real – and will probably get worse.

Time, April–May 2000, page 61

Global warming is causing more extreme weather including floods and droughts.

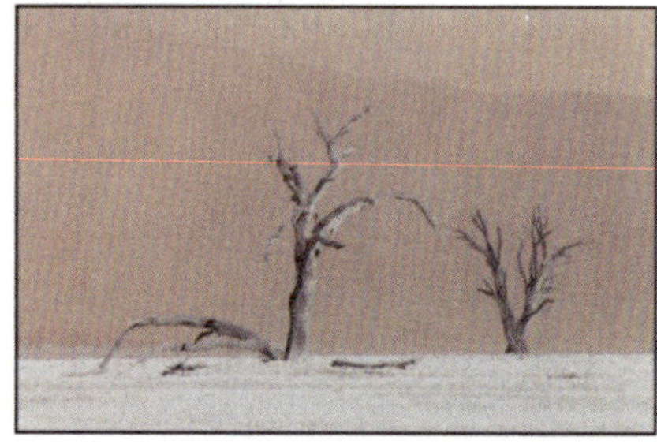

Global warming killing thousands already

Global warming may have killed up to 100,000 people in the past three years, and threatens to cause mass migration, disease, poverty and even war.

Scientists generally agree that the Earth's atmosphere is heating up, probably as a result of carbon dioxide emissions causing the 'greenhouse effect'. The warming is disrupting weather systems, resulting in more extreme weather, including droughts, torrential rain and hurricanes.

NZ Herald, 24 July 2000 reprinted from *Observer*

The evidence

Scientists are able to be fairly confident about the five global warming conclusions on page 42 because they are based on large amounts of collected factual data on climate. The graph on average global surface temperatures and the CO_2 data are just two examples.

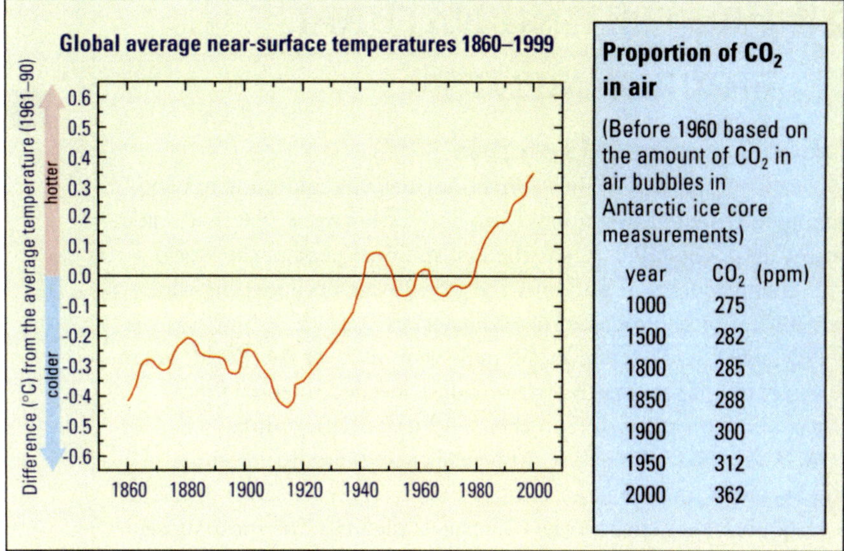

Proportion of CO_2 in air

(Before 1960 based on the amount of CO_2 in air bubbles in Antarctic ice core measurements)

year	CO_2 (ppm)
1000	275
1500	282
1800	285
1850	288
1900	300
1950	312
2000	362

Global warming predictions

Climate changes are still not well understood and their future effects are very difficult to predict. However, using the data they already have, scientists design computer models of how the climate may change in the future, under different conditions. Here are a couple of predictions:

By 2050 (most of you will still be alive!):

1 **Average global temperatures will be up by 1 to 1.5°C** (the range of predictions vary from 0.5°C to 2°C).

2 **Average sea level will rise by 20 cm** (the range of predictions varies from 5 cm to 40 cm).

It is hard to predict how this will affect different places. However these sorts of things will certainly happen (some already are happening!):

- Some low-lying parts of the world will be under water – such as some Pacific Islands and the coast of low-lying places such as Bangladesh.
- Some places will be wetter, others drier.
- Warm climate crops will be able to be grown nearer the poles.
- Pests and diseases of the tropics will move into areas that are currently cooler.
- Some plants and animals of cold places may become extinct.

What can be done?

New Zealand and many other countries (but not all), have signed an agreement (Convention on Climate Change) to reduce the release of greenhouse gases. Governments now must make sure this happens.

Activities: Global warming examined

1 Evidence about global warming:

A From the global temperature graph write down what has happened in the last 20 years.

B From the carbon dioxide levels data:

a Plot this data on a line graph.

b Explain how measuring the proportion of CO_2 in an air bubble in ice can give an accurate picture of CO2 levels 1000 years ago.

c Describe the trend shown by the graph.

d Explain why this graph suggests that the changes may be related to human activity. (Clue – population graph page 4.)

2 Analysing newspaper reports

a From the *Time* report list three facts that the article uses as evidence that human activity is affecting global warming.

b How does the *NZ Herald* article link the deaths of 100,000 people to global warming?

3 Understanding the process

Study the greenhouse effect diagram

a Look up and explain these terms:

> infrared radiation, troposphere, fossil fuels, greenhouse gases (list three)

b Write down why increased carbon dioxide levels cause the atmosphere to warm up.

4 The effect of humans

In 1997 the NZ Ministry for the Environment estimated that increased greenhouse gases caused by human activities were in the following proportions:

Carbon dioxide	36%
Methane	45%
Nitrous oxide	19%

a Draw a pie diagram showing these proportions.

b For each of the three gases explain what human activity is involved.

c Explain why planting trees would help reduce the amount of CO_2 in the air.

d List three other ways that humans can reduce the release of greenhouse gases.

5 Discussing future problems

In your groups discuss these problems:

a The poor countries say: 'The wealthy countries are causing these problems but we are the ones that suffer.' Do you think this is a fair argument? Why?

b The USA produces about one third of all greenhouse gases released by humans, yet it has not been willing to sign the agreement to reduce these gases. Why do you think this is the case, and what should be done?

24 Protecting natural environments

National parks and reserves

How can we protect our most important natural environments and endangered plants and animals? In 1872 Yellowstone National Park was set up in the USA. It was the first national park in the world. Since then governments all over the world have been setting aside special areas to be protected in various ways.

National parks operate in different ways all over the world. Some are so protected that special permission is needed to enter them. Others allow people to live in them and graze their animals in them. A few (including those in New Zealand) are free to enter. Others charge tourists large sums of money. (It costs about $NZ200 to get a permit to land on the Galapagos Islands.) The most special of the world's national parks have been named World Heritage Sites.

Marine reserves

In recent years it has become apparent that in many places sea life is also under threat, and attempts are being made to set up marine reserves. These are areas of sea, seabed and shoreline which have been set aside for scientific research, public access and enjoyment, and where marine life is totally protected. No fishing or gathering of marine life is allowed. Although New Zealand has 13 marine reserves, they cover less than one per cent of the coastline and most have only recently been set up.

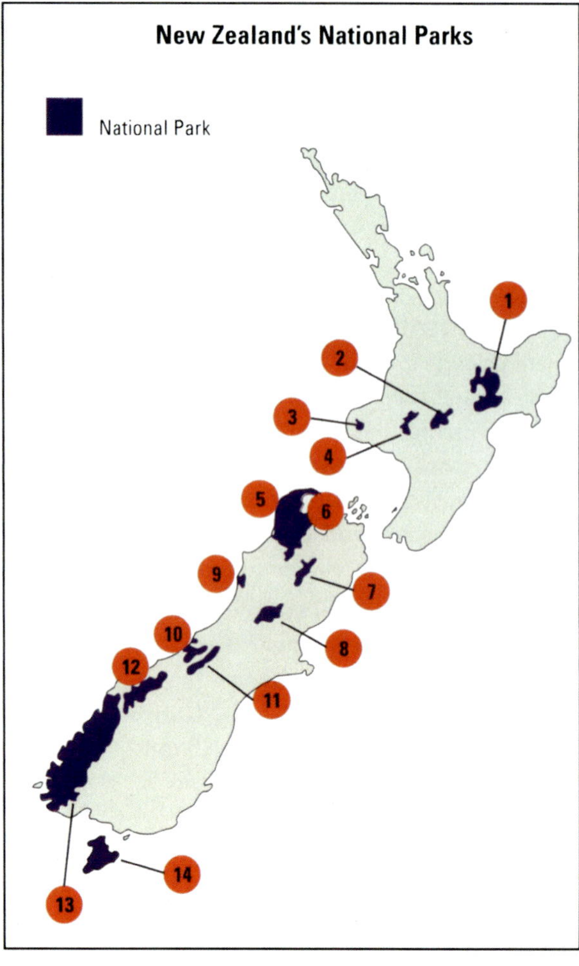

New Zealand's first national park was set up in 1894 when the Tongariro area was given to the people of New Zealand by the local Maori people. New Zealand now has 14 national parks, 19 forest parks, and many smaller reserves and maritime parks which together cover over 16 per cent of the country. The 14 National Parks are:

a	Abel Tasman	b	Arthur's Pass
c	Egmont	d	Fiordland
e	Kahurangi	f	Mount Aspiring
g	Mount Cook	h	Nelson Lakes
i	Paparoa	j	Tongariro
k	Urewera	l	Westland
m	Whanganui	n	Rakiura (Stewart Island)

Tourists and national parks

Setting up and running national parks and reserves may seem quite straight forward operations. But they have many problems. **Tourists** in large numbers may damage the parks and overcrowding may spoil the experience.

A new kind of tourism has grown up. This is **ecotourism**. Ecotourists are people who wish to visit remote places and experience natural environments without harming them. They will often pay large amounts of money to do this. But even ecotourism may have a price. No matter how careful ecotourists are, some environments are just too fragile to avoid being damaged if too many people visit them.

Case study: Leigh Marine Reserve – north of Auckland

This marine reserve was set up in 1975. Although it is quite small, extending only a kilometre or so along the coast, it was quite a battle to set up. Many local fishermen see marine reserves as ways to lock up the coastline and prevent them from fishing or gathering shellfish wherever they wish. However, it has now become clear that marine reserves become places where marine life can breed and populate other areas. They are also important for scientific research and for people to visit and enjoy. In fact, this reserve has become a real treasure of the sea. Swimmers or snorkellers find that they are surrounded by hundreds of fish and that the rocks are covered with seaweed, kina, shellfish and other marine life.

The success of this small reserve is now becoming a problem. Over 100,000 people visit it every year. It is great that so many people want to visit an unspoilt natural environment, but the sheer numbers of visitors are likely to pose a threat to the survival of the marine life. Should controls be put on who can visit it?

National parks in developing countries

Generally, in wealthier countries the parks are well looked after and supported, but in poorer countries it is a different story. The government is less likely to have the money to set up protected areas. Most people will have little understanding of biosecurity issues and the need to save endangered species. In some countries with rapidly growing populations there is a shortage of farmland and it is difficult for local people to be convinced that shutting up a habitat to save a rare species is more important than developing it for farming. Some parks that are already set up are under pressure from farmers to move into them and farm them. The rare Mountain Gorillas of the Congo in Africa are in serious danger for this reason. Protecting African rhinos and elephants from poachers who can sell the horns and tusks for large amounts of money is another huge problem.

Case study: Ranomafana National Park – Madagascar

Madagascar is an extremely poor country. Many of its people have had to live by subsistence farming. This involves burning the forests and growing crops for a few years until the soil is exhausted, and then moving on to destroy more forest. Unfortunately this destroys the habitat of the many very unusual plants and animals – such as lemurs, which are now endangered species. Madagascar has not got the money to set up national parks to protect them. The setting up of Ranomafana National Park is an example of what is happening in several poor, undeveloped countries.

When a new species of lemur was discovered in 1986 an American University provided money and worked with the Malagasy Government to set up the Ranomafana National Park in 1991. This park covers 43,000 ha of forested hill country and provides a protected home for 12 species of endangered lemurs, as well as for many rare birds and reptiles. The park has been set up to employ local people, about 80 of whom are earning money by working in the park, many as guides. Although Ranomafana National Park is in a very remote place and has only been operating for a few years, it is already beginning to attract tourists. In 1999 over 10,000 ecotourists travelled over rough roads in difficult conditions and paid a large fee to see the rare animals. Half of this money goes to improve the education and health of the people who live near the park, and to run programmes to help them make their farming sustainable.

Activities: National park issues

1 New Zealand's national parks

Match up the names of the 14 national parks to the numbers shown on the map (you may need to check your atlas).

2 The value of national parks

Imagine attempts are being made to set up a new national park in a valuable wilderness area. You have been asked to plan a one minute TV ad to persuade people to support it. Outline the commentary and images you would show.

3 Case study: Leigh Marine Reserve

a List three benefits a marine reserve may have.

b Why do people object to marine reserves?

c Discuss in your group the ways that could be used to prevent the marine reserve from being 'loved to death'. Write your conclusions down.

4 Case study: Ranomafana

a Why was the park set up?

b Who organised and paid for setting it up?

c At first the local people were very suspicious of the forest being shut up for a park. Explain their worries.

d List three ways in which the local people are benefiting from the setting up of the park.

e Explain one advantage of the tourists coming to the park.

f Explain one possible problem that tourists may cause in the future.

5 Making decisions

The management of national parks involves making some difficult decisions. Imagine your group have to make management decisions for a national park. Here are different viewpoints on two management issues. Discuss the answers and decide what decisions you would make:

A Should visitors be charged to visit a national park?

a Yes. The costs of looking after the park should be paid by those who visit.

b No. The parks belong to everyone and the government should pay for their upkeep.

B Should parks have more tracks and roads to make access easier?

a Yes. The easier it is for people to use them the more they will be used and enjoyed.

b No. If it is made too easy to use the parks they will be damaged by too many people, and pests will get in. The parks should only be for those really keen to walk in them and to look after them.

25 The future of the Earth is in our hands

In what shape is the Earth?

Here is a summary of the state of the Earth at the beginning of the new millennium:

The bad news

1 The population of the Earth is increasing at an alarming rate.
2 The resources of the Earth are being put under extreme pressure to supply the wants and needs of the expanding population of humans. Water, energy and marine resources are under particular pressure.
3 The natural environments of the Earth are under pressure. Habitats are being destroyed, species are being lost and replaced by less valuable species.
4 The wastes that humans make are causing big pollution problems.
5 Global warming is beginning to change conditions.

The good news

1 Population growth is beginning to slow very slightly.
2 People are gradually becoming aware of the pressure that we are putting on our habitats and other resources.
3 More work is being done to find and develop sustainable energy sources and better farming practices.
4 Waste reduction schemes are becoming more widely used.
5 Governments around the world are at least talking about ways of reducing the harmful effects of human activities and in some cases are taking some action.

The next fifty years

Remember that most of you will still be alive in 50 years time. The predictions are that the Earth will have to cope with up to 3 billion more people by then. The bad news box shows that we already have big environmental problems now. The good news box shows that although we are beginning to work on these problems, we are not yet doing enough. We are certainly not doing enough to cope with the needs of another 3 billion people! So what must be done?

What can we do?

Most of these problems are so big that they can only be solved by the actions of governments and big businesses. But protecting the environment of the Earth is not just a problem for governments, it is everyone's problem.

Governments' actions follow what the people who elect them want. Business will become environmentally friendly if their customers demand it. So we as individuals can make a difference. We need to do two things:

A Be informed:

1 Find out about environmental issues. Do research.
2 Find out what other people think. Get different points of view. Be objective- look at facts. Do not get carried away by emotions.
3 Weigh up information and make informed decisions about what should be done.

B Be involved:

Do not just sit back and let others deal with issues.

1 At home, be environmentally aware:
   - Save energy, save water.
   - Recycle wastes when you can.
   - Recycle organic matter.
2 Get involved in your local environment – tree planting, clean ups, etc.
3 Get involved with local conservation groups.
4 Make your voice heard. Let the decision-makers know your views – write, phone, e-mail, visit them.

Further research ideas

The future of the earth is in our hands. Being effectively involved with environmental issues means that you need to be able to use a range of skills – research, presentation, communication.

You will already have used some of these skills in doing some of the investigations earlier in this book. Some other environmental issues that you could investigate are:

1 The ozone hole issue (see page 47)
Possible ways of tackling this issue are outlined in detail as a guide for working on other issues.

2 Recycling
Look at the issues involved in recycling items such as different kinds of scrap metal. Find out how it is collected, how it is processed, how the recycled material is marketed, what costs are involved, etc.

3 Marine reserves
What is involved in investigating if a new one is needed near you, how to sample public opinion, what is involved in preparing and presenting a case to the Department of Conservation.

4 Antarctica
This continent is very fragile. Can it remain protected? Investigate the competition to use its resources and the attempts being made to protect it.

Activity: Investigating an environmental issue

The ozone hole

The presence of the ozone hole has recently become an important environmental issue. It is specially worrying for New Zealanders. Do a detailed investigation of this issue. (You or your group could tackle the whole issue, or parts could be shared by different people or groups):

A Find out what the problem is

This will involve using your research skills. You may get information from the library, websites, or by questioning experts. Get answers to the following questions:

1 What is ozone, and what does it do in the air?
2 What is the ozone hole and where is it?
3 How big is it and how does it change with the seasons?
4 How has it changed over the last 20 years?
5 Why is it a problem to humans?
6 What seems to be causing it?
7 How is it measured?

B Present your data to others

Prepare publicity material about the ozone hole – either as a poster, or as ideas for a TV or film presentation. You need to show in a simple way:

1 What the ozone hole is
2 Why it is a problem for us
3 What we can do about it.

C Sample public opinion about dealing with the effects of the ozone hole

Design a simple questionnaire about the issue. Design about 10 questions to find out what people know about the ozone hole and its effects and what they feel they would do as protection from its effects (if anything). Sample opinion by giving the questionnaire to others and analysing the results.

D Decide what should be done about it

Discuss the issue with others in your group, then list the actions (if any) that need to be taken to solve the problem by:

1 Individuals
2 Government
3 Any other affected groups.

Some thoughts to ponder on

The following are quotations about the environmental problems of the Earth. In your groups:

- decide what its real environmental message is, and how true it is.
- discuss if it should affect the way we think and live.

A 'Nature does not care if we live or die. We can not survive without the oceans, for example, but they can do just fine without us.'
(Quote from *Time*, April–May 2000, page 15.)

B 'There will be 150 million environmental refugees by 2050.'
(Environmental refugees are people who will be displaced by the effects of global warming, as a result of climate changes, and especially the rise in sea level.)

C 'It takes five hectares to supply the needs of one person in the affluent world. If all the poor people used that much it would take five Earths to provide for them.'
(The calculation is based on what it takes to supply all the wants of people in wealthy countries.)

D 'Live on this earth as if it is forever.'
(Quotation from Reith lecture, 2000.)

E Discuss the six important statements of fact about the earth and the pressures on it that are in italics in this book. They are on pages 2, 7, 20, 24, 40 and 42.

Some environmental groups

These are active in New Zealand and also have websites, etc. with useful information:

- Royal NZ Forest & Bird Protection Society
- Greenpeace
- Maruia Society (forest protection)
- Local Conservation Groups

Sources of environmental information

These have websites with useful information:

- World Wide Fund for Nature New Zealand (WWF)
- Department of Conservation
- Ministry for the Environment

Index

Glossary of Technical Terms

birth rate – the number of children born per woman.

browser – animal that feeds on leaves, twigs and other parts of plants.

ecosystem – a community of interacting organisms and the environment in which they live.

endemic – a species of living things that is only found in one area.

exotic – plant or animal that was introduced or arrived from abroad. Not indigenous or native.

fossil fuels – fuels made from the remains of living things that have been compressed in the earth (coal, oil, natural gas).

greenhouse gases – gases in the atmosphere that trap heat (the greenhouse effect), mostly carbon dioxide, methane, nitrous oxide.

habitat – the place where a plant or animal lives.

incinerate – to destroy by burning.

indigenous – living things which are native to the place in which they are living.

infrared – radiation from sun that is strongly absorbed and reflected.

leachate – underground polluted water.

migration – movement of living things from one place to another.

non-renewable resources – resources that once used up are no more. They cannot be replaced.

population growth rate – the rate at which a population grows when the number of deaths is taken from the number of births in a given time.

quarantine – kept in isolation, away from other living things.

toxic – poisonous.

troposphere – lower part of Earth's atmosphere, extending about 10 km up from the Earth's surface.

ultraviolet (UV) – rays from the sun that burn the skin.

(Some other terms are explained on page 3)